THE ENERGY OF SLAVES

THE ENERGY

Andrew Nikiforuk

THE ENERGY OF SLAVES

OIL AND THE NEW SERVITUDE

David Suzuki Foundation

GREYSTONE BOOKS
D&M PUBLISHERS INC.
Vancouver/Toronto/Berkeley

12 13 14 15 16 5 4 3 2

Greystone Books
An imprint of D&M Publishers Inc.
2323 Quebec Street, Suite 201
Vancouver BC Canada V5T 4S7
www.greystonebooks.com

David Suzuki Foundation
219–2211 West 4th Avenue
Vancouver BC Canada V6K 4S2

Cataloguing data available from Library and Archives Canada
ISBN 978-1-55365-978-5 (cloth)
ISBN 978-1-55365-979-2 (ebook)

Editing by Barbara Pulling
Copyediting by Stephanie Fysh
Jacket design by Peter Cocking
Text design by Heather Pringle
Jacket photograph © Jason Reed/Getty Images
Printed and bound in Canada by Friesens
Distributed in the U.S. by Publishers Group West

We gratefully acknowledge the financial support of the Canada Council for the Arts, the British Columbia Arts Council, the Province of British Columbia through the Book Publishing Tax Credit, and the Government of Canada through the Canada Book Fund for our publishing activities.

Greystone Books is committed to reducing the consumption of old-growth forests in the books it publishes. This book is one step towards that goal.

Contents

. . . . .

To the masters and the slaves

"When civilized, as well as barbarous nations, have been found, through a long succession of ages, uniformly to concur in the same customs, there seems to arise a presumption, that such customs are not only eminently useful, but are funded also on the principles of justice... To ascertain the truth therefore, where two such opposite sources of argument occur; where the force of custom pleads strongly on the one hand, and the feelings of humanity on the other; is a matter of much importance, as the dignity of human nature is concerned, and the rights and liberties of mankind will be involved in its discussion."

THOMAS CLARKSON, *An Essay on the Slavery and Commerce of the Human Species*, 1785

"Freedom is the concern of the oppressed."

ALBERT CAMUS

Prologue

. . . . .

THE GODS OF ancient Greece possessed frightful energies. Zeus, the father of them all, could fling thunderbolts. Helios directed the sunlight; Ares ruled war; Demeter guided growing plants; and moody Poseidon often shook the earth or raised the restless sea. The Greek gods could fly through the air, change the course of rivers, place constellations in the sky, and even cloak themselves with invisibility.

The ends to which the gods put their great powers were not always lofty. They often spent their fearsome energy to settle quarrels, commit adultery, or simply overcome boredom. But their exploits filled ordinary Greeks with wonder. Many dreamed of living like the gods. The Olympians commanded endless power at their fingertips; the immortals did not sweat like slaves or strain like oxen.

Yet the profligate energy-spenders on Mount Olympus had deep concerns about sharing power with lesser beings. When it came time to make humans and the earth, the gods assigned the task to two Titans: clumsy Epimetheus ("afterthought") and his clever brother Prometheus ("foresight"). Epimetheus made animals first. He gave them speed, claws, fur, and enormous strength. But that didn't leave much for people. Seeing that human beings had been left naked and shoeless, Prometheus acted quickly. He stole fire from the forge of kind Hephaestus and the mechanical arts from Athena, the goddess of wisdom. In so doing, Prometheus brought energy and technical innovation to early civilization. With fire, humans could now smelt metals and even harden ceramics. "And in this way man was given the means of life," relates Plato in *Protagoras*, a minor dialogue.

Zeus, of course, did not take kindly to this theft. He feared that Prometheus might steal more energy and so punished him cruelly: he chained the Titan to a rock and ordered an eagle to pluck out his liver anew every day. Many people believe the parable ends there. They think that poor Prometheus liberated humans from darkness, then got punished for sharing power and innovation. But that's not the whole story.

Plato continues the yarn in *Protagoras*, where it crackles with consequences. With the gift from Prometheus, people changed their lives dramatically. Thanks to fire, they began to cook food, heat their dwellings, destroy forests, and build grand cities. But soon they were using this transformative gift unwisely. Their numbers grew without control, and they waged incessant war. Destruction reigned on earth. Fearing that mortals might exterminate themselves, Zeus finally intervened. To end the excess, he ordered Hermes to give humans two moderating gifts: justice *(dike)* and respect *(aidos)*. When winged Hermes asked if he should offer these gifts to a few mortals or the many, Zeus instructed that the gifts be shared among all people, "for cities cannot exist, if a few share only in the virtues." His reasoning was plainly divine: if all people did not discern the consequences of employing energy and its technologies, hubris would again engulf human affairs.

Although this parable originates more than 2,500 years ago, it retains its freshness. As noted by the French philosopher Fabrice Flipo, the story of fire tells us that no energy is clean, free, or unlimited and that the use of every Promethean tool must be carefully measured. Only justice and respect can defeat excess. In the absence of proportion and scale, energy invites destruction and dispersion. And it can create a servitude that blinds master and slave alike.

Petroleum companies and petrostate leaders champion the first part of this fable. They claim that they have bettered civilization with fossil fuels and have created a second, if not greater, Promethean revolution. A new fire, fed by mineral fuels, has enabled the development of labor-saving machinery: energy slaves. The proliferation of these inanimate slaves dependent on coal or oil has changed every facet of life. Ordinary people in geographies blessed with oil now possess the power of Greek gods. Using their mechanical slaves, they can fly through the air, blast off mountaintops, grow the population of cities, drain rivers, and even cause earthquakes. What the petro fable doesn't tell us is that accelerated development has also depleted the gift of coal and oil. And embedded in this unprecedented power is the deeply problematic relationship between master and slave.

This second Promethean revolution did not draw its values from fire. It built upon the institution of human slavery, which once served as the globe's dominant energy institution. Both Aristotle and Plato described slavery as necessary and expedient. We regard our new hydrocarbon servants with the same pragmatism. To many of us, our current spending of fossil fuels appears as morally correct as did human slavery to the Romans or the Atlantic slave trade to seventeenth-century British businessmen.

But in the absence of wisdom and temperance, every energy relationship becomes a matter of dominion rather than stewardship. Petroleum's Olympian impact has caused human civilization to again fall prey to hubris. Once dependent on the energy of slaves, we are now slaves to petroleum and its masters. And this time, without Zeus to rescue us, we must challenge the ancient paradigm ourselves and find our own path to using energy on a moral, just, and truly human scale.

1

The Energy of Slaves

. . . . .

"Men grow reconciled to everything, and even to servitude, if not aggravated by the severity of the master."

MONTESQUIEU, *The Spirit of Laws*, 1748

AT THE END of the twentieth century, Donella Meadows, a pioneering environmental scientist, pondered the nature of energy servitude. The organic gardener and lead author of the 1972 bestseller *The Limits to Growth*, a critique of high-energy living, admitted that, yes, she was a damned slave owner. Although her energy servants came with names like High Octane and Black Gold, she possessed chattel—just like Thomas Jefferson, the author of the Declaration of Independence. Her assertion stunned and angered many Americans. How could a man as respected and intelligent as Jefferson have owned slaves and still call himself a democrat? What on earth was Meadows thinking?

Meadows' admission reflected honestly how energy molds our lives: Jefferson employed scores of slaves at his Virginia plantation and had a child with one of them; Meadows burned up to thirty barrels of oil a year to get her work done. Jefferson relied on the energy system his contemporary Thomas Paine once called "Man-stealing." Thanks to petroleum, Meadows employed invisible slaves bound with chains of carbon that came from ancient plants.

Yet Meadows did not believe that either consuming oil or owning slaves implied a blind acceptance of an energy system. Jefferson, who inherited hundreds of slaves, called the energy institution of slavery "the opprobrium of infidel powers." Although he once wrote that certain people were probably born to be slaves, he defined slavery as "this great political and moral evil." Yet, as Meadows pointed out, Jefferson knew that in an agrarian society, giving up slavery would mean abandoning his livelihood, his place in the culture, and his beloved plantation, Monticello. It would mean surrendering both power and comfort, something he wasn't willing to do.

Donella Meadows understood this reality too. The energy sources and energy converters adopted by any society ultimately become that society's cultural and political masters. Energy and its abundance or scarcity order the day. "I believe that wasting energy, emitting greenhouse gases, buying and discarding stuff mindlessly, driving species to extinction is as immoral as keeping slaves," Meadows wrote. "Yet I drive a car and fly on planes. Mountains of chewed-up trees flow through my life in the form of paper. There's polyvinyl chloride, a substance I would love to ban from the earth, all over my house." So she did her best to use her petroleum slaves sparingly and "to reduce [her] load on the groaning earth." Meadows knew that every form of energy consumption, from slavery to oil, involves, somewhere, a sacrifice.

BEFORE COAL AND OIL, civilization ran on a two-cycle engine: the energy of solar-fed crops and the energy of slaves. Shackled human muscle built, powered, and emboldened empires from Mesopotamia to Mexico. The ancients, our practical ancestors, understood the cost and the laws of energy. Slaves made efficient energy converters and created healthy surpluses. On a minimum of calories provided by cereal crops, a group of organized slaves could, well, move mountains—carry a litter of rich people, build irrigation works, fight wars, or simply make life easier for their masters.

Most ancient civilizations (with the exception of the Chinese, with their unique agro-energy system) mobilized the muscle of slaves with no moral qualms. When Hammurabi, the king of Babylon, established the world's first legal codes two thousand years ago, he decreed death to any person who helped or sheltered escaping slaves. The Greeks called their slaves *andrapoda,* or man-footed creatures. Many aboriginal societies around the world used captives as domestic servants. The prophet Mohammed had slaves too. But no other society employed slaves as greedily as the Romans.

Captive human muscle was the fuel that drove, expanded, and ultimately undermined the Roman Empire. In 200 BC, the rising power boasted nearly 600,000 slaves. Two centuries later the numbers had swelled to several million, with anywhere from a quarter to a third of the population at any given time consisting of slaves captured in war or acquired through trade. In some jurisdictions slaves even outnumbered freemen. Slaves so dominated Roman life that the Senate adroitly defeated a proposal to put slaves in uniform to distinguish them from poor citizens. Such a provision, the Roman philosopher Seneca explained, would simply illustrate "how great would be the impending danger if our slaves should begin to count our number."

Romans, like most other ancients, considered slavery an ugly but necessary business. They divided the world into two distinct classes: the *domini* (the masters of energy) and the *servi* (the providers of energy). The ancients could draw only on certain finite options: solar-powered plant energy, wind, animals, and the brute strength of humans. The agronomist Varro defined slaves as speaking instruments: "Tools are either endowed with speech, half speech or silence. To the category endowed with speech, the slaves belong, to that of half speech the oxen, to the silent the carts." Masters generally used their "speaking tools" as they pleased. Slaves had no legal rights and could not marry. Their offspring belonged to the master. Masters could sell, buy, and loan slaves; they could whip, chain, or even torture them.

A good slave lasted only about twenty years before he or she died or was freed, so the Roman empire needed about half a million new slaves every year. At first the empire procured these high-energy providers with ease, primarily through conquest or by kidnapping free men and women on the road or the seas. At one point the illegal pirate trade in slaves trafficked as many as 10,000 people a day. Orphaned children and convicted criminals played a prominent role in swelling the energy ranks, and many citizens sold themselves into slavery to pay their debts. In 209 BC, after the pillage of Carthage, Rome put 130,000 slaves on the market. A war in Judea replenished the Roman economy with 70,000 individuals. Julius Caesar (a virtual state-owned energy company) may have sent back as many as half a million slaves from his conquests.

Some Roman slaves worked for the state, repairing aquifers, roads, and temples, but most toiled for private citizens. At first the majority worked on large rural estates, raising crops, milling grain, or herding livestock. As the empire's cities grew richer, more slaves worked as artisans or domestic servants. Some worked as letter carriers, accountants, doctors, or cooks; many were trained in literature and copy writing. Personal slaves cut hair, prepared baths, and performed sexual services for their masters. Wealthy urban Romans even walked the streets with a *nomenclator*, a slave who remembered the names of important people. Dwarves and disfigured slaves entertained, and fit slaves shed blood as gladiators. Slaves lit the lamps in the morning and dampened the hearth-warming fires in the evening. Agricultural slaves envied the lighter, specialized loads of their city cousins.

Like every subsequent energy innovation, Roman slavery began with tentative respect and ended with careless waste. At first rural landowners employed slaves as part of the household, in which the slaves were often treated almost as equals. But as estates grew and slaves became ever cheaper, the relationship changed. In his famous essay *The Spirit of Laws*, the eighteenth-century French social critic Montesquieu described the change this way: "But when the Romans aggrandised themselves; when their slaves were no longer the companions of their labour, but the instruments of their luxury and pride; as they then wanted morals, they had need of laws. It was even necessary for these laws to be of the most terrible kind." One law ordained that the murder of a master by a servant warranted the death of every slave in the household.

On the island of Delos, then a Saudi Arabia of muscle, ten thousand slaves could be sold in a day. Slave traders hid defects and fattened slaves for sale. Each marketable slave was stripped naked and came with a scroll proclaiming his or her character, weaknesses, and origin. A false account allowed the buyer to return his labor-saving instrument within six months. Boys, eunuchs, doctors, and beautiful women typically fetched high prices.

Ordinary Roman freemen might own one slave, while members of Rome's middle class might employ ten. Wealthy landowners owned hundreds, even thousands. The slave of a slave (important slaves had them) was called a *vicarius*. Romans typically greeted each other with *Quot pascit servos?* ("How many slaves do you feed?"). The answer revealed a person's status, wealth, and access to energy.

Romans calculated the calories burned by slaves with an economic coolness. Cato the Elder offered precise advice on how to properly maintain a slave: "For those who work in the fields, during the winter, for modii of triticum wheat, during the summer four and half modii; for shackled slaves, during the winter four pounds of bread, when they begin to hoe the vine, five pounds of bread until figs begin to ripen; then return to four pounds... To add to the bread of slaves, keep a stock of as many fallen olives as you can, and given them hallec [salted fish] and vinegar."

Some Romans did speak out against the excesses of slavery. The wealthy slaveholder and stoic Seneca penned many epistles that read like reflections on how to use energy wisely and fairly. For example, Seneca didn't consider the feeding of a slave to a pond full of ravenous lamprey for breaking a goblet (an actual incident) a prudent form of energy management. He advised that only "dumb animals" required the lash. The Spanish-born patrician suggested that respect produced better slave energy than fear.

Seneca, who later committed suicide on orders from the Emperor Nero, also worried that the institution of slavery had conferred godlike powers upon many masters. With too much energy at their disposal, wealthy Romans had become fat and indolent. In one famous epistle, Seneca describes a rich master dining in front of a retinue of slaves. One slave had been trained to cut fat capons, another to pick up tipsy guests, and another to serve wine while dressed as a woman. "The master eats more than he can hold, and with monstrous greed loads his belly until it is stretched and at length ceases to do the work of a belly; so that he is at greater pains to discharge all the food than he was to stuff it down," wrote Seneca. "All this time the poor slaves may not move their lips, even to speak. The slightest murmur is repressed by the rod; even a chance sound,—a cough, a sneeze, or a hiccup,—is visited with the lash. There is a grievous penalty for the slightest breach of silence. All night long they must stand about, hungry and dumb. The result of it all is that these slaves, who may not talk in their master's presence, talk about their master."

Seneca never opposed slavery outright, and he vigorously supported the status quo. The noble controlled energy; the ignoble furnished it. But he recognized that the empire's slave system had undermined Rome's traditional values: self-reliance, endurance, resilience, strength, and honor. "We are driven into wild rage by our luxurious lives," he wrote, "so that whatever does not answer our whims arouses our anger. We don the temper of kings. For they, too, forgetful alike of their own strength and of other men's weakness, grow white-hot with rage, as if they had received an injury, when they are entirely protected from danger of such injury by their exalted station. They are not unaware that this is true, but by finding fault they seize upon opportunities to do harm; they insist that they have received injuries, in order that they may inflict them."

Perhaps the most interesting observations about Roman energy consumption came from Epictetus, the slave philosopher. After obtaining his freedom, Epictetus spent the rest of his life thinking about the right way to live. His writing appeared at a time when energy use in the empire had peaked and slaves were getting harder to acquire. To Epictetus, true liberty meant not being a thoughtless consumer: "For freedom is acquired not by the full possession of the things which are desired, but by removing the desire. And that you may know that this is true, as you have laboured for those things, so transfer your labour to these; be vigilant for the purpose of acquiring an opinion which will make you free; pay court to a philosopher instead of to a rich old man." Or, as he later put it, "Give me a man who cares how he shall do anything, not for the obtaining of a thing but who cares about his own energy."

The availability of cheap slave labor not only changed Rome's masters; it also discouraged energy innovation. The Romans, good organizers and diligent builders, eschewed science and new technologies. With so many slaves available to do the work, the water mill and the harvester got scant attention. As French historians Jean-Claude Debeir, Jean-Paul Deléage, and Daniel Hémery later concluded, "The social incentive to develop machines powered by other sources of energy than humans was weak or non-existent." The Dutch raconteur Hendrik Willem van Loon even came up with an explanatory law: "The amount of mechanical development will always be in inverse ratio to the number of slaves at a country's disposal."

Romans complained about their slaves routinely and often thought of them as enemies. Runaways, liars, and thieves were branded on the forehead. Escaped slaves, called *fugitivi*, represented lost property and lost income and also posed a general menace to the public. Bands of runaway slaves typically became brigands. A legendary bandit known as Bulla Felix once sent this pointed message to authorities, along with a captured centurion: "Give this message to the masters: Pay your slaves their maintenance wages in order that they may not become brigands." Organized businesses hunted down runaways and returned the fugitive energy sources to their owners. Moses Finley, a historian of ancient slavery, has observed that "fugitive slaves are almost an obsession in the sources."

As Rome expanded through Europe and Africa, the empire acquired enormous energy surpluses that it had trouble controlling. Many of those captured as slaves were former soldiers. From 133 to 72 BC, four major slave revolts unsettled Rome. Romans regarded the revolts with fearful disdain. Livy, the famous Roman historian, condemned slave revolts as particularly wicked events and recommended that they be dealt with "not solely in that spirit which we use toward other enemies" but with a focused fury.

The Servile Wars exposed the importance of slavery as a power source. In Sicily, home to large estates energized by speaking tools, tens of thousands of slaves revolted. One insurrection there took authorities four years to put down. The great Spartacus rebellion involved nearly seventy thousand armed slaves and paralyzed the empire for almost a year. Every revolt created a temporary energy crisis, delivering a political shock as it deprived the Roman economy of essential manpower. The crucifixion of six thousand slaves after the Spartacus rebellion reminded citizens that Rome *would* sacrifice energy to maintain its energy system.

The anthropologist Joseph Tainter and other historians note that Rome's dependence on slave energy created a predatory economy: "Defeated peoples provided the economic basis, and some of the manpower, for further expansion. It was a strategy with high economic returns." Whenever the empire needed more energy and cash, it recruited yeoman farmers from the countryside for service in the legions and sent them on campaigns of annexation. Rural landowners scooped up the lands abandoned by these peasants and purchased slaves to energize their expanding estates. Triumphant armies returned to Rome with tens of thousands of slaves. As Debeir, Deléage, and Hémery note, "Every military adventure was an opportunity to transfer property to the advantage of the urban aristocracy, to augment the weight of servile labour in the civilian economy, which clamoured for more slaves and therefore ever more war."

For several centuries this energy system served the empire well. What Rome could not achieve locally, it gained through geography. Instead of seeking intensive energy returns, it focused on exploiting surpluses from a widespread area: Africa and the Mediterranean. With large supplies of slaves flooding the marketplace, the average Roman paid no domestic taxes. Stolen gold and a tax on slaves even kept the state running. But soon the empire ran out of easy geography to annex and encountered increasingly belligerent slaves (the Germans and the Celts). As a consequence the empire's energy returns and economic surpluses gradually dwindled. Guarding borders and provisioning a large army depleted state coffers.

Roman leaders responded to the chaos with simplistic solutions. In response to civil wars and invading tribes, the emperor Diocletian tried a new recipe: a larger army, a devalued currency, and a larger bureaucracy for collecting higher taxes. In its early years, the Roman Empire had employed 130,000 in its army; toward the time of its demise, it kept 650,000 men under arms. Yet not even these numbers could keep the barbarians out. Nor could higher taxes replace the earlier surpluses generated by the energy of slaves. And so began Rome's long collapse. "Overall it was a considerable increase in complexity," reflects Joseph Tainter. "Transforming the Roman Empire into a sort of an early version of the Soviet Union was a solution—of a kind—that retarded collapse of a couple of centuries but that, in a certain way, made it unavoidable. The Roman Empire could not afford such a large army and, eventually, it destroyed itself in the attempt of maintaining it."

As Rome dissolved, so did its major energy system: slavery. When the empire drew in its borders and began collecting fewer slaves, the price of human chattel rose. The cost of slaves, combined with diminishing returns from rural estates due to soil erosion, created a series of crises. Large estates fell idle or were underworked. Slaves and free peasants often supported barbarian invaders. The wealthy abandoned city life and turned their estates into self-sustaining communities. As the economy collapsed, it became cheaper to employ freemen than to hold slaves. Ultimately, more slaves became serfs, whom the Romans considered "slaves of the land itself to which they were born."

AFTER THE COLLAPSE of the Roman Empire, slavery endured in fits and starts throughout Europe and the Middle East. The English word *slave* comes from the post-Roman practice of enslaving Slavic peoples. But slavery no longer existed as a singular energy monopoly. During the Middle Ages energy innovations such as water mills and wind power, along with small landholdings, made slavery less appealing. But the ugly renewable energy system did not disappear: it reemerged in the sixteenth century when Europeans conquered the New World. After Old World diseases such as smallpox emptied much of the continent of its aboriginal people, the conquerors realized they didn't have the manpower to mine the place. Confronted with an energy shortage, the Europeans turned to Africa, a bountiful muscle reservoir, and created one of the world's most sinister energy systems: the Atlantic slave trade. Just about every European nation participated, but by the seventeenth century the British were easily dominating the business. As many as 90 wind-powered slave ships carried

35,000 slaves across the Atlantic in the late 1750s alone. With millions of imported slaves, the Europeans mined gold and established vast sugar, indigo, rice, coffee, tobacco, and cotton plantations.

The concept of free labor made as much sense to the rich and powerful in the Age of Empire as renewable energy and reduced energy consumption do to elites today. How would England eat and clothe itself without slaves working in plantations in the Caribbean? Arthur Young, a British agricultural critic and statistician, noted in 1772 that only 33 million of the world's 775 million people actually lived in freedom. Servitude under monarchies remained the global norm. The world's energy system, dependent on human muscle, demanded centralized control and organization. One famous American essayist, William Goode, defended the portability of Britain's colonial energy system in a typical argument: "The labor of the slave like every thing else, will go where it is most useful—will meet the most effectual demand."

The Atlantic slave trade functioned like an hourglass. It collected energy over a vast African terrain, concentrated and channeled the supply of captives for transport, then sold the human energy at a wide number of destinations to power agricultural enterprises. The monarchies of West Africa, long accustomed to employing slaves in armies, in gold mines, and on farms, captured and transported the "human machines" to forts along the coast. There, slave traders from England, Portugal, Spain, Holland, and France bartered textiles and trinkets for the shackled energy. Once crammed onto two-hundred- to five-hundred-ton vessels, the slaves began their perilous journey in blood and excreta to the New World, where their muscle power energized Brazil, Cuba, the West Indies, and the United States. Due to the appallingly high death rate in transport and on sugar plantations, the demand for slaves seemed insatiable. It was also profitable. Slave states came to depend on the energy as a source of easy revenue. Portugal, France, Spain, and Britain secured immense fortunes from the taxes on the highly volatile business of slave trading. The British government was soon subsidizing the deadly exchange with forts, soldiers, and naval bases.

The Atlantic slave trade wasted energy like a leaking pipeline. Over a three-hundred-year period, as many as 10 million Africans crossed the Atlantic in chains. For every successful import, two to five corpses littered the high seas or African slave ports. A slave ship could be identified from the gangs of sharks in its wake. The American historian and civil rights activist W.E.B. Du Bois estimated that the American slave trade "meant the elimination of at least 60,000,000 Negroes from their fatherland." Modern scholars put the final toll between 22 million and 55 million. Du Bois defined slavery as a system in which war captives did the work so warriors could capture more energy. The trade corrupted Central Africa. "Whole regions were depopulated, whole tribes disappeared; villages were built in caves and on hills or in forest fastnesses; the character of peoples like those of Benin developed their worst excesses of cruelty instead of the already flourishing arts of peace," he wrote. The trade, he added, bled the continent of its energy and spirit.

The slave trade, as an energy system, compromised the whole of European society. Every lump of sugar dropped into British teacups depended on shackled muscle. John Locke, the noted philosopher of liberty, held shares in the Royal African Company, which branded all of its property with hot irons. Voltaire mocked the brutal treatment of slaves but didn't object to having a slave ship named after him. A slave plantation in Barbados helped fund a library for Oxford's All Souls College. Even well-known philanthropists traced their fortunes to the slave trade. Edward Gibbon, the historian, used the proceeds from his grandfather's slaving business to write *The Decline and Fall of the Roman Empire*, observing that slaves, "an unhappy condition of men," "endured the weight, without sharing the benefits, of society." The Church of England owned a plantation, and a popular evangelist, George Whitefield, maintained that "hot countries [could] not be cultivated without Negroes." Businessmen, of course, regarded the energy traffic as "the hinge on which all the trade of this globe moves." Slaves, the New World's first energy boom, were cheaper than freemen.

Like their Roman counterparts, slave traders and slave owners lived opulently. The slave traders of Nantes, for example, sent their dirty laundry all the way to Haiti to be scrubbed in mountain streams, because Haiti's water turned everything whiter than Brittany's. Caribbean slave owners were especially notorious for their extravagance. Saint-Domingue (later Haiti), an island of forty thousand Europeans and half a million slaves, offered the masters an endless menu of distractions: orchestras, gambling dens, and even a traveling wax museum. The life of French plantation owners was, as visitors reported, "divided between the bath, the table, the toilette and the lover."

Europeans generally treated their plantation slaves much worse than the Romans had. Plantation economies operated out of sight and were run by absentee landlords. But like the Romans, the British and other slavers took a dim view of unplanned energy shortages. Planters religiously hunted down fugitive slaves and enacted strict laws on the return of runaways. In the United States, slaveholders hired night guards and funded private militias to prevent energy leaks in the system. Most Caribbean planters also reasoned that it was cheaper to wear out a slave and buy a new one than to let the slaves breed. Children acted as a drain on adult productivity.

Alexis de Tocqueville, a French aristocrat and brilliant social observer, visited the United States in 1831 and was struck by the impact of slavery on people's habits and character. The manufacturing North depended on coal and free labor, while the South drew its wealth from the energy of slaves—tobacco and cotton demanded unremitting muscle. The well-read Tocqueville recognized thirty years before the Civil War that slavery would become a defining life and death issue for the U.S. South.

Tocqueville described most Northerners, who had grown up without slaves or servants, as self-sufficient and enterprising, if not "patient, reflecting, tolerant, slow to act, and persevering in [their] designs." In contrast, slaves supplied the "immediate wants of life" for many Southerners. "The American of the South is fond of grandeur, luxury, and renown, of gayety, of pleasure, and above all of idleness; nothing obliges him to exert himself in order to subsist." While the Northerner placed the pursuit of comfort and wealth "above all the pleasures of the intellect or the heart," Southerners eagerly spent the prosperity supplied by the energy of nearly four million slaves on military games, pleasure, and excitement. "The citizen of the South," he observed, "is more given to act upon impulse; he is more clever, more frank, more generous, more intellectual, and more brilliant." Northerners possessed all the good and evil aptitudes of the middle class, while Southerners possessed the tastes and prejudices "of all aristocracies." Slavery, in short, diminished "the spirit of enterprise amongst the whites."

Tocqueville was writing mostly about the antics of large slaveholders. The surplus generated by their massive plantations founded beautiful cities such as Richmond and Savannah. But the majority of the South's 400,000 slaveholders belonged to a diverse middle class. They owned not hundreds of slaves but five or fewer. Many small slaveholders were professionals or shopkeepers who owned slaves to generate extra wealth or improve their upward mobility. The more energy a small agricultural slaveholder could deploy in the fields, the faster he could climb in Southern society.

Many such slaveholders cared little for elaborate homes or fancy clothing, because they were always on the move. After exhausting soils in one region, the masters picked up their slaves and families and sought new landscapes to mine. When Frederick Law Olmsted, the U.S. journalist and landscape designer, visited a Texas slaveholding family in the late 1850s, he found nothing but ugliness. The slaveholders lived in a miserable shack, but they commanded ten field hands who earned them an enviable annual income. "What do people living in this style do with so much money?" Olmsted asked. "They buy more negroes and enlarge their plantations."

But some middle-class slave owners recognized the sins of their occupation. Many subscribed to an evangelical Protestantism that "sparked an intense fear that they were unworthy of the prosperity born of slavery and that their children would be unable to resist the materialism that was so much part of the market culture," writes James Oakes in *The Ruling Race*, an insightful history of American slaveholders. Struck by remorse and guilt, they debated the perils of commanding an energy institution that generated easy wealth, and in particular they worried about the institution's impact on their service to God. They feared that too much power and prosperity would undermine their resilience and spoil their children. "In this state of slavery I almost feel that every apparent blessing is attended with a curse," wrote one master. Explained a Virginian slaveholder to a Northern friend, "You may feel very happy that you are not in a slave state with your fine Boys, for it is a wretched country to destroy the morals of youth." As Oakes notes, many masters of black energy felt trapped in a dehumanizing system they had not made. "We were born under the institution and cannot now change or abolish it," opined one Mississippi slaveholder. Occasionally, a master would free his slaves due to a "sense of right, choosing poverty with a good conscience, in preference to all the treasures of the world." But the benefits of commanding surplus energy won most practical debates. Exclaimed an Alabama slave master in 1835, "If we do commit a *sin* owning slaves, it is certainly one which is attended with *great conveniences*."

It would take a new energy master to unshackle this order of human slavery. Although fossil fuels at first promised widespread liberty and virgin utopias, they eventually delivered something different: an army of fuel-hungry mechanical workers that would require increasingly complex forms of management and an aggressive class of powerful carbon traders. Without much thought, we replaced the ancient energy of human slaves with a new servitude, powered by fossil fuels.

2

Slaves to Energy

. . . . .

"The fact is that civilization requires slaves. The Greeks were quite right there. Unless there are slaves to do the ugly horrible uninteresting work, culture and contemplation become almost impossible. Human slavery is wrong, insecure and demoralizing. On mechanical slavery, on the slavery of the machine, the future of the world depends."

OSCAR WILDE, *"The Soul of Man under Socialism,"* 1891

ALFRED NORTH WHITEHEAD, the popular English philosopher and mathematician, once wondered how the abolition of something as cruel and unjust as human slavery could have taken so long to achieve. The riddle confounded historians, theologians, and even abolitionists. Why did several thousand years elapse before an organized antislavery movement appeared in the eighteenth century? Why did philosophers as great as Aristotle and Seneca have no objections to fettering human muscle? Why did Christians challenge the excesses of slaveholders but not the institution itself? Why did Hindu, Hebrew, Islamic, and African law all approve of and accept slavery?

In 1933, Whitehead offered a tentative answer: "It may be impossible to conceive a reorganization of society adequate for the removal of some admitted evil without destroying the social organization and civilization which depends on it." He then added a caveat: "An allied plea is that there is no known way of removing the evil without the introduction of worse evils of some other type."

Whitehead's assessment wasn't far off the mark. Mechanical slaves, powered first by coal and later by oil, effectively eliminated the need for widespread human slavery and serfdom. The new carbon-based slaves did not immediately replace human ones—in many cases, they made conditions worse for decades—but they did change human thinking. They also made the muscle power of slavery seem old-fashioned, in the same way automobiles made horses look quaint. In so doing, the hydrocarbon age created a new class of masters and a unique form of energy servitude. And this new inanimate order presented distinct challenges and inequities, on a geographic scale never before experienced. Societies once powered by human muscle, animals, sun, and sail now boarded locomotives. Combustion engines devoured hydrocarbons mined from the ground. The rapid adoption of fossil fuels and their labor-saving devices altered the course of civilization.

England didn't pioneer the burning of coal for heat; the Chinese did. But England did industrialize the messy hydrocarbon to get more work done. Having felled most of their hardwood forests for cooking and heating, the fourteenth-century British faced a genuine wood scarcity. So they turned to a stinky substitute: coal. Demand grew persistently, and by the eighteenth century coal miners constituted a singular class of men. The British aristocracy regarded miners with the same contempt Roman elites had had for their slaves, and conditions in the coal mines were as onerous and as dangerous as those on sugar plantations.

Even with this dedicated workforce in place, coal mining dug up a challenge. As water started to flood some deeper mines and drown miners, the resource became harder to get. Early engineers resorted to buckets, hand pumps, and horses to haul out the coal. But horses cost money to feed, so the nation looked for a cheaper alternative. Ironmonger Thomas Newcomen first fashioned a "fire engine," a one-piston device that ran on coal. Shortly afterwards, James Watt, a quiet Scottish mechanic, improved the design of Newcomen's mechanical pump and vastly improved its efficiency. Watt's refined model for the steam engine, a fancy tool to mine more coal, appeared in 1769 and was in full production in 1775. It was no coincidence that Thomas Clarkson, the great English anti-slavery leader, launched his campaign for abolition just twelve years later. The poet Samuel Coleridge tellingly called Clarkson "a Moral Steam-Engine."

Alfred René Ubbelohde, a Belgium-born physicist, argued in 1955 that slavery probably prevented the invention of the steam engine some 1,700 years before its British arrival. Both pistons and the forceful properties of steam were known to ancients, but given healthy economic returns, slaveholders weren't interested in alternative technologies. Their apathy delivered "incalculable consequences" for world history, says Ubbelohde. "The economic incentive for developing the inanimate power was neutralized by facile harnessing of animate energy [slaves] in the ancient world."

But when the new technology finally appeared, the power generated by the steam engine made slavery redundant.

These heat engines pumped water, ground corn, sawed lumber, crushed oil from seeds, formed lead into sheets, and twisted ropes and cables. By 1827 tanneries, iron foundries, dockyards, and breweries all employed these stupendous inventions. In his *Treatise on the Steam Engine,* British engineer John Farey claimed that one machine with the force of 120 horsepower was equal "to the strength of more than a thousand men acting together." Just 750 people in a cotton mill powered by steam engines could spin as much thread as 200,000 people unassisted by mechanical slaves. Moreover, steam vessels navigated rivers "much more steadily than any galleys can be rowed by slaves."

Charles Fourier, a French radical and utopian socialist, viewed this multiplication of inanimate slaves as a profound threat to human labor. In the early nineteenth century Fourier estimated that the amount of work performed by "various mechanical inventions" in Great Britain alone equaled that from "*two hundred millions* to *four hundred millions* of working adults." He described this unwearying army as "slaves patient, obedient, submissive; from whom no rebellion need be feared" and warned that "we shall see masters engaging, as the cheapest, most docile and least troublesome help, the machine instead of the man."

Andrew Carnegie, the U.S. steel magnate and author of *The Gospel of Wealth,* offered more astonishing numbers on the horsepower revolution in 1905. Watt had created the unique form of measurement based on the energy expended by one horse to raise a bucket of coal out of a mine. In his reverential biography of Watt, Carnegie described the power of the machine: "One horse-power raises ten tons a height of twelve inches per minute. Working eight hours, this is about 5,000 tons daily, or twelve times a man's work, and as the engine never tires, and can be run constantly, it follows that each horse-power it can exert equals thirty-six men's work; but, allowing for stoppages, let us say thirty men."

In 1824, England's steam engine puffed out 26,000 horsepower, the equivalent of nearly 750,000 men or 100,000 horses. The powerful steam machine was first adopted by new industries such as ironworks and large-scale cotton production. It became ubiquitous wherever coal could be mined cheaply. The first steamship popped up in 1801, followed by the first locomotive in 1814. Within thirty years, Watt's improved model had changed the energy fortunes of industrial England and the world. When King George III asked industrialist Matthew Boulton why he built steam engines, the keen innovator replied, "I'm engaged, your Majesty, in the production of a commodity which is the desire of kings." When the king asked what that was, Boulton answered, "Power, your Majesty." By the 1880s, the output of the world's steam engines totaled 150 million horsepower. Running but a sixteen-hour shift, these machines collectively exerted the work of more than 3 billion humans. With the world population at slightly more than 1 billion and only one in five of those a mature male, Carnegie wasn't the only one to appreciate the unprecedented revolution at hand. In just one hundred years, coal-fired machines added 3 billion invisible slaves to the global economy.

And one of the most celebrated children of this radical new surplus of power was abolition. When James Watt began his tinkerings, few British political or intellectual leaders questioned the slave trade. Economist Adam Smith reckoned that "the love of dominion and authority over others will probably make [slavery] perpetual." European merchants named their slave ships *Equality* and *Liberty* with no sense of irony. Yet by 1838 England's most powerful social protest movement had ended the reign of one of the world's oldest energy systems throughout the British Empire. Reluctantly, the United States, Cuba, and Brazil followed suit some thirty years later.

The scale and effectiveness of the abolition movement shocked slave traders and slave owners alike. The movement broke all convention and precedent, argues historian Adam Hochschild in *Bury the Chains*. Abolition's appearance in eighteenth-century England was, he says, as if a sudden global movement to ban automobiles emerged today. "For reasons of global warming, air quality, noise, a dependence on oil, one can argue, the world might be better off without cars," he writes. "And what happens when India and China have as many cars per capita as the United States? Even if you depend on driving to work, it's possible to agree there's a problem. A handful of dedicated environmentalists try to practice what they preach, and travel only by train, bus, bicycle, or foot. Yet, does anyone advocate a movement to ban automobiles from the face of the earth?" But that's exactly what the abolitionists did with slavery. And the new energy slaves paved the way. As a pioneer in securing work from peat and coal, as well as sail, England possessed an imperial surplus of energy. It could afford the luxury of anti-slavery. Non-coal-adopters could not—yet. "In no non-western countries did abolition emerge independently as official state policy," notes slave-trade historian David Eltis. "And no non-western intellectual tradition showed signs of questioning slavery per se."

The great anti-slavery debates of the eighteenth and nineteenth centuries sound oddly familiar today. The slavers angrily played the economic card. British traders in Liverpool, for example, argued that they employed thousands of sailors and porters. Moreover, the system made slaves "happy" and prevented the overpopulation of Africa. If the British didn't carry on the slave trade, then the French would take over, went the defense. Some acknowledged the basic ugliness of the business but added that "neither was the trade of a butcher an amiable trade, and yet a mutton chop, was, nevertheless, a good thing." In an 1836 report, Brazilian slave owners demanded to know how the world could function without slaves: "What would become of America's export trade? Who would work the mines? The fields? Carry on the coastal trade?"

Powerful slave-trading lobby groups in Britain also invoked cognitive dissonance. No tobacco smoker or tea drinker who added sugar to their favorite beverage, let alone any cotton consumer, had the right to question an economy highly dependent on slaves. American slaveholders employed similar arguments. They called Europe's half-starved industrial workmen a "little experiment" and "a cruel failure." In 1831 a Virginia legislator argued that the energy of slaves was irreplaceable: "There is slave property of the value of $100.000.000 in the State of Virginia, &c., and it matters but little how you destroy it, whether by the slow process of the cautious practitioner, or with the frightful dispatch of the self-confident quack; when it is gone, no matter how, the deed will be done, and Virginia will be a desert." Without slavery, added another proponent, the civility and equality of white populations would erode: "Freedom is not possible without slavery." One slave lobby group, sounding like modern spin doctors, suggested simply changing the name of the business: "Instead of SLAVES, let the Negroes be called ASSISTANT-PLANTERS." Others proposed the term "servants."

But abolitionists challenged these platitudes by using the new steam-powered press. With the help of graphic posters, pamphlets, and illustrations, the movement made visible the casual brutality and high mortality of the slave trade for slaves and sailors alike. Distributed by train and steamship, these tracts depicted severed limbs, shackled children, sickened sailors, and naked men. Thomas Clarkson, a tireless and well-informed crusader, highlighted the crowding on ships in particular—"chained like herrings in a barrel"—as well as the trade's appalling bloodiness. Other new political tools were just as effective. Abolitionists flooded the British parliament with petitions. Hundreds of thousands of women denounced tea as a "blood sweetened beverage" and boycotted sugar grown by slaves. The capitalist Josiah Wedgwood created a design of a kneeling man in chains with the motto "AM I NOT A MAN AND A BROTHER?" The jasperware cameo, which later became a medallion and hair ornament, served as one of the movement's most popular fashion statements.

Countering the claims of the slave-trade lobby, William Wilberforce, the great parliamentarian, gently emphasized that England's tea drinkers "ought all to plead guilty." The Quakers, who had at one time engaged in the dirty business themselves, argued that "to live in ease and plenty by the toil of those whom violence and cruelty have put in our power" violated both Christianity and common justice. The ever-observant Montesquieu argued that "ingenious machines" could ultimately do the work of slaves.

Abolition seeded or presaged a series of related social upheavals. Once England had outlawed slavery, citizens redirected their political energies toward ending child labor, championing union rights, and promoting women's suffrage. Even the earliest of feminist treatises drew the parallel: "Is one half of the human species," asked Mary Wollstonecraft, author of *A Vindication of the Rights of Women,* "like the poor African slaves, to be subject to prejudices which brutalize them?" Even the U.S. civil rights movement owed a debt to the success of the abolition crusade. And each of these important political battles reflected the quantitative increases in energy that transformed working relationships by putting more mechanical slaves into the marketplace. In each case a subordinate class challenged a master class and fossil fuels magnified political liberties. Each of these successful movements was the direct result of cheap and high returns from fossil fuels and their machines.

WHILE ABOLITIONISTS successfully railed against the ball and chain, scientists and engineers hailed the steam engine as an impressive auxiliary to human might. The expansive powers of steam, they declared, had in fact changed humankind's role on earth. Thanks to the heat transferred from coal to machine, the lowly human could become the "Lord of Creation." In 1831, the Boston Society for the Diffusion of Useful Knowledge printed the first volume of its new American Library of Useful Knowledge: a collection of previously published essays and extracts on the new mechanical world order. In it, an 1829 lecture by Daniel Webster praises the machine as a convenience booster, a work reducer, and a toil mitigator. Machines, Webster says, will stretch "the dominion of mind, farther and farther, over the elements of nature, and by making those elements, themselves, submit to human rule, follow human bidding, and work together for human happiness." An extract by John Herschel, a famous English scientist and astronomer, calculated "the virtue" in a bushel of coal and said it could transcend "whole nations of men." Based on the performance of Cornwall steam engines, Herschel determined that a device burning a bushel of coal could lift 70 million pounds of material a foot high. A man's daily labor was equivalent to the burning of 4 pounds of coal. Two pounds of coal could, then, lift a man to the summit of Mont Blanc without any human toil. The Great Pyramid of Egypt, 12,670 million pounds of (Herschel believed) granite, could be raised by the effort of about 630 chaldrons of coal, "a quantity consumed in some founderies in a week." (A chaldron, sometimes "cauldron," was about 36 heaping bushels of coal.) The annual consumption of coal in London—1,500,000 chaldrons—could raise a "cubical block of marble, 2200 feet in the side" 8 miles high. Concluded Herschel, "The inherent power of fuel is, of necessity, greatly underrated."

Charles Babbage also grasped the magnitude of the transition unfolding in England. The engineer—a banker's son and the inventor of the first computer (called a "calculating engine") and other marvels—offered insights on the new energy regime in his 1832 book, *On Economy and the Machine and Manufacturers.* "The discovery of the expansive power of steam, its condensation, and the doctrine of latent heat," he noted, "has already added to this small island, millions of hands."

Mechanical slaves, enthused Babbage, not only surpassed human power but reduced the time required to do things and converted worthless substances into valuable products. He differentiated between two kinds of machines: those that made power (the steam engine) and those that executed work. Babbage, who detested "Mobs," street music, and "Negro slavery," praised the uniformity and steadiness of mechanical slaves. His book identified the workers toiling with machines as "men" and the machine owners as "masters." Babbage recognized that there might be limits to the "rapidly increasing power" of mechanical slaves, but he believed that his own thinking machines could temper civilization's "ever accelerating force."

Not everyone appreciated this acceleration of life. Thomas Carlyle, a leading critic of the Industrial Revolution, warned that mechanistic thinking would cause upheavals and ruin.

Nor did he consider factory conditions and urban slums much of an improvement on slavery. In one famous essay, "Signs of the Times," the aristocratic historian and essayist thundered against the Age of Machinery: "Men are grown mechanical in head and in heart, as well as in hand. They have lost faith in individual endeavour, and in natural force, of any kind. Not for internal perfection, but for external combinations and arrangements, for institutions, constitutions—for Mechanism of one sort or other, do they hope and struggle. Their whole efforts, attachments, opinions, turn on mechanism, and are of a mechanical character."

The energy system dispatched by the British abolition movement had been largely employed in distant colonies, but American abolitionists relentlessly attacked their own Southern cousins, often using explosive rhetoric that championed the rise of the machine. In 1853, U.S. educator Horace Mann declared, "Had God intended the work of the world should be done by human bones and sinews, he would have given us an arm as solid and strong as the shaft of a steam engine; and enabled us to stand, day and night, and turn the crank of a steamship while sailing from Liverpool or Calcutta... Ignorant slaves stand upon a coal mine, and to them it is only a worthless part of the inanimate earth. An educated man uses the same mine to print a million of books." In many respects the U.S. Civil War marked the first of the world's several convulsions that brought high-energy societies into direct conflict with lower-energy or less mechanical ones. "The North defeated the South by a campaign of attrition supported by coal mines, steel mills and railroads," writes energy historian Earl Cook. "The South was clearly superior in military talent and social determination, but it could not prevail against the inexorably grinding superior power machine of the North." The conflict, which killed 400,000 people and cost $40 billion, proved that energy transitions do not come cheaply. Nor did the wrenching conflict "create a new energy base for the south," writes American sociologist Fred Cottrell. Until the 1940s, Southerners preserved a culture "little different in energy terms from that of Egypt under the Pharaohs," alongside the greatest concentration of mechanical slaves in the world. It would take the discovery of oil in Louisiana, Oklahoma, and Texas to change that.

3

The Oil Pioneer

. . . . .

I once was unknown by the happy and gay
And the friends that I sought did turn away...
But now what a change! Our house is so grand,
No one is so fine throughout the whole land.
And we can now live in the very best style,
And it's simply because my "Pa has struck ile."

POPULAR AMERICAN SONG, 1860s

WHEN THE United States married petroleum, the nation matured into a carbon-rich fairy tale. The Oklahoma Corporation Commission (OCC), which has regulated oil drilling in that state for nearly one hundred years, tells the story in a blunt PowerPoint presentation it calls a "History of Energy." Oil supplies 40 percent of U.S. energy needs and "keeps our country moving," declares the OCC. "Almost our entire transportation fleet—our cars, trucks, trains and airplanes—depend on fuels made from oil. Lubricants made from oil keep the machinery in our factories running. The fertilizer we use to grow our food is made from oil. We make plastics from oil. It is quite likely that the toothbrush you used this morning, the plastic bottle that holds your milk, and the plastic ink pen that you write or draw with are all made from oil." Without black gold, says the commission, America wouldn't be America. Or free.

The presentation then asks its school audience to conjure up a fantastic vision: "Imagine a lake 10 miles long, 9 miles wide and 60 feet deep. Fill that lake with oil. That would be about as much oil as the entire world uses in one year. The United States would use about 1/4 of it." The commission hints that there might be problems ahead but doesn't elaborate. Imagine, for example, this great lake of oil getting smaller, thicker, and costlier by the day. Now imagine most of the lake has actually moved offshore, to foreign countries whose names many Americans don't know or can't spell. Then imagine more countries demanding a bigger share of the lake to live the American dream—a vision Americans are progressively losing due to the high cost of extreme or imported oil. The pioneer of oil has grown fat slurping his own petroleum frontiers and now holds a straw but no milkshake.

The Age of Petroleum began in the United States with fire, mayhem, and the promise of easy wealth. Men like Henry R. Rouse pioneered the trade. While the "clear headed and prosperous" New York lumber dealer was felling the wilds of western Pennsylvania in 1859, he caught scent of a "rock oil boom" in the region. A former dry-goods clerk known as "Crazy Drake" had drilled an oil well at Titusville, and "the news flew like a Dakota cyclone." The oil, refined into kerosene, made a cheaper lamp fuel than pricey whale oil.

Henry Rouse headed for the oil regions. He wasn't alone. It was the nation's biggest adventure since the California gold rush. According to early oil historian J.T. Henry, the boisterous cast of characters included "anxious drillers, the smiling, wealthy fortunates, the downcast, ruined unfortunates, the busy teams conveying the barreled liquid to the water, the oil-begrimed and mud-besmeared boats, the eager barterer and the earnest seller." At the beginning of the boom, Rouse bought vast tracts of land near a place called Enterprise. Like Drake, he soon struck crude, and "wealth poured in upon him in fabulous volume." But the market was volatile. Rouse sold oil sometimes at $14 a barrel, other times at 40 cents. It all depended on the number of new holes and the number of refiners.

Had Rouse lived, according to Henry, he could have been either "a giant or a bankrupt in the oil business." But we'll never know. On the evening of April 17, 1861, one of Rouse's wells flowed giant amounts of liquid crude. Men scrambled for barrels to capture the wealth as it poured from the ground. Then came a sudden conflagration as leaking natural gas found an open flame. Two wells, a tank, and a barn burst into a ball of fire. Burning men ran from the inferno. One witness compared the running workers to "a rapid succession of shots from an immense Roman candle."

Henry Rouse was only twenty feet from the well when it exploded. He threw his wallet and a book of valuable papers outside the line of fire then ran for a ravine but fell. By the time he got out of the petroleum furnace, half of his body had been burned to a crisp. Rouse maintained consciousness for an hour before he died. He dictated his last will and testament, including a stipend for his father, $100 to the two men who carried him out of the fire, and half of his princely fortune on Well No. 8 to the poor of Warren County.

The well burned for three days. Millions of drops of oil spouted off in one direction, "presenting all the hues of the rainbow making a scene like enchantment." Nineteen men had lost their lives, and another ten or so bore disfigurements for life. The United States started a global energy revolution with blood, dirt, spills, and chaos. But the enduring boom would transform the country into the world's "illuminator and lubricator." Oil, as chronicler John J. McLaurin put it, would "dispel gloom and chase hobgoblins."

Kerosene, oil's first triumph, secured its initial hold on American life with substantial government backing. At the time, Americans could choose from a range of liquid fuels to light up their homes, including whale oil (the rich man's fuel), camphene (turpentine mixed with alcohol), lard, coal oils (smoky kerosene), and candles. To pay for the costs of the Civil War, the government heavily taxed best-selling camphene, at $2 a gallon. That gave kerosene, taxed at only 10 cents a gallon, an unfair advantage. "The petroleum industry did not arise in response to market conditions but rather in response to government intervention," explains journalist Bill Kovarik. "The petroleum industry was born with the silver spoon of subsidy wedged firmly in its teeth."

Rock-oil developers also exercised a Yankee gift for self-promotion. The kerosene crowd claimed that their product saved whales, when in truth it was the rising cost of chasing fewer creatures that was doing that. In the 1880s, the *Baltimore Merchants' and Manufacturers' Directory* hailed petroleum as "promotive of human civilization and happiness"; oil gladdened the cabins of pioneers, lit the huts of miners, and "cheered the home of the thrifty farmer." In just 20 years, petroleum production had grown from 500,000 barrels to 25 million. "Its bright rays lend their kindly aid to the thousand homely cares which give zest and happiness to the family circle," continued the merchants.

"Thus the sum of human knowledge is increased and the aggregation of wealth added to by the useful occupancy of hours snatched from darkness and sleep."

In 1896, John J. McLaurin, a vivacious oil salesman, described the success of the "grandest industry of the ages" in his *Sketches in Crude Oil*. Nothing kept a steam engine running faster and cleaner than petroleum, McLaurin wrote. The Galena Oil-Works, which he said ranked among "the most noteworthy representative industries of Uncle Sam's splendid domain" (and which was a Standard Oil subsidiary), offered catchy jingles about its products on the home front, too:

> Have you a somewhat cranky wife,
> Whose temper's apt to broil?
> To ease the matrimonial strife
> Just lubricate when trouble's rife—
> Pour on Galena Oil.

OIL ALSO HELPED to change the fresh face of American capitalism. With oil, John D. Rockefeller, a thrifty Baptist bookkeeper with a knack for numbers, established a new business model for the country's economy. Business was no longer about making a living. It was about making a killing, and the surplus capital created by oil made that possible. The savvy Rockefeller avoided the messy business of wildcatting and invested instead in refining, transporting, and distributing kerosene, creating one of the world's first multinational corporations, Standard Oil. In short order, he demonstrated one of petroleum's most singular attributes: its ability to concentrate power and generate capital.

In building his monopoly, Rockefeller pioneered several critical business practices. He set up a statistics department to keep track of costs and prices. Staff ate lunch together and talked about money and numbers. The board of directors met every day. The company also strove to create a standard kerosene product that didn't immolate innocent consumers. (Fires caused by bad batches of kerosene dispatched as many as eight thousand U.S. citizens a year in the 1870s.) With a personal promise to "expose as little surface as possible," Rockefeller introduced an abiding code of secrecy into the industry. And he committed his firm to the outright destruction of all competitors.

Rockefeller's monopolistic tools included espionage, bribery, bullying, blackmail, sabotage, and drawbacks. By controlling costs for tankers, barrels, and other parts of the business, he systematically undercut the competition. Whenever independent refiners refused to sell, Standard dropped the price in the marketplace to create a "good sweating." The oil titan helped to set up the South Improvement Company, which negotiated a secret rebate deal with railway companies forcing competitors to pay double what it cost Standard to move oil. The company bribed politicians and paid individuals to monkey-wrench uncooperative competitors. By 1880, Standard Oil controlled nearly 90 percent of the refining business and kerosene had become the United States' fourth-largest export. It was even being sold to China.

Rockefeller's ruthlessness toward small businesses started an "oil war." Public outcry and angry newspaper articles described the company as a "monster" and "an anaconda." Through it all, Rockefeller quietly maintained that "it is not the business of the public to change our private contracts." Oil companies from BP to ExxonMobil still live by the Rockefeller creed.

It was Standard's rude concentration of power that gave birth to the muckraking tradition in American journalism. In 1903, Ida Minerva Tarbell, the daughter of an oil refiner put out of business by Rockefeller, wrote a 19-part exposé of the company for *McClure's Magazine* and published it the next year in book form: the 850-page investigative masterpiece *The History of the Standard Oil Company*. Tarbell documented how the nascent oil business had turned a forest into a $200 million global marketplace that supported 60,000 people. "But suddenly," she wrote, "at the very heyday of its confidence, a big hand reached out from nobody knew where, to seal their conquest and throttle their future. The suddenness and blackness of the assault on their business stirred... their sense of fair play and the whole region arose in a revolt which is scarcely paralleled in the commercial history of the United States." Tarbell's meticulous work, which documented the unethical nature of big business, haunts the highly concentrated industry to this day. (Nearly a hundred years later, a retired Shell Oil Company executive would write a book called *Why We Hate the Oil Companies*. Like Tarbell, he singled out self-interest.)

Standard's money made banks and changed landscapes. One of Rockefeller's partners, Henry Flagler, took his share of the Standard Oil fortune in the 1890s and developed Florida. Flagler turned swamps into tourist destinations, built fancy hotels, and constructed railways that later brought Americans by the millions to Miami and West Palm Beach. The elderly oil tycoon even built a marble palace full of gold for his thirty-two-year-old third wife. Finally, in 1911, the Supreme Court of the United States ordered the breakup of Standard Oil into thirty-four separate companies. Standard Oil of California became Chevron, Standard Oil of New Jersey morphed into Exxon, and Standard Oil of New York grew into Mobil. Together with BP, Royal Dutch Shell, Gulf, and Texaco, these private firms would dominate global oil markets between

1940 and 1970. All behaved like Standard Oil, setting the code for the state oil companies that would eventually control the majority of the world's oil industry.

Just as electricity threatened to diminish the kerosene market, Henry Ford came to the rescue in 1908 with his Model T. Before that, "horseless carriages," whether electric or powered by an internal combustion engine, had been playthings for the rich. Ford, who purposely made a product his workers could afford, democratized mobility and created a hot market for previously unsellable gasoline. Refiners had considered the flammable product a troublesome waste product of kerosene distillation and dumped it into rivers. At one point the Cuyahoga River in Ohio contained so much gasoline that steamboats often set the water on fire.

Henry Ford did with the automobile what Rockefeller had done with oil. He standardized the vehicle, mechanized its manufacturing, and concentrated the industry. Most early Ford buyers lived in farming communities in the north-central states, but the highest sales numbers occurred in oil-boom towns. Nothing changed the United States faster or more completely. The automobile promised to "renew the youth of the overworked man or woman" with a mobility that was, as one Oldsmobile advertisement put it, "always ready when you are—a race horse when you want speed." Soon every major U.S. city complained about hellish traffic. In 1935, the Roosevelt administration estimated that the United States commanded 930 million kilowatts of installed power and that motor vehicles accounted for more than three-quarters of that energy.

Novelist William Faulkner would later remark that "the American really loves nothing but his automobile: not his wife, his child nor his country nor even his bank-account first." Two Texas historians, Roger Olien and Diana Davids Olien, reckon that the turn-of-the-century revolt against Standard Oil was driven partly by panic at the demise of the American ideal of manhood. Founders of the republic such as Jefferson, Paine, and Franklin had envisioned an agrarian nation run by sun, muscle, and slaves, one whose citizens prized above all independence. They foresaw a country that shunned great wealth, valued quality craftsmanship, and championed self-sufficiency. Fossil fuels celebrated a different set of values. The mechanical slaves of oil and coal rewarded bigness, quantity, and concentration. Oil in particular promised fast money and a life of idleness. Petroleum accelerated industrial changes that reduced the number of self-employed men from 88 percent of the population in 1860 to one-third by 1910. "This change, making men economically dependent on others, was a calamity for masculinity," write the Oliens in *Oil and Ideology*.

It didn't take long for oil, combined with the marvels of electricity and coal, to create a distinct new political economy, too. Thorstein Veblen, a quirky economist of Norwegian heritage who spoke twenty-six languages, took advantage of Rockefeller's generous endowment to the University of Chicago to write a critique of America's new energy culture: *The Theory of the Leisure Class*. By 1899 Veblen was seeing the first marks of wasteful energy spending on a people in "the vicarious consumption of goods." He foresaw the emergence of hydrocarbon's many fruits, including the evolution of technology and science into the dominant shapers of everyday life. And by the time Veblen speculated on an oil stock in old age and lost his entire investment, he had witnessed oil's transformation of the United States into the richest country on the planet. In 1920, 45 million citizens worked in the growing oil economy under industrial masters who paid out some

$77 billion in wages. Mark Twain, who occasionally lunched at Standard Oil offices, often reflected on oil's impervious hold on his country. "Like all the other nations, we worship money and the possessors of it," he said. "We have the two Roman conditions: stupendous wealth, with its inevitable corruptions and moral blight, and the corn and oil pensions—that is to say, vote-bribes, which have taken away the pride of thousands of tempted men and turned them into willing alms-receivers and unashamed."

Oil petrolized American geography wherever it was discovered, but nowhere so much as in Texas. In 1901, a gusher at Spindletop, near the town of Beaumont, roared for ten days and spilled a million barrels of oil before it was contained. Speculators and wildcatters invaded the field. Derricks sprouted like weeds and men slept in town on billiard tables. The rich field launched four major oil companies—Gulf, Texas, Humble, and Magnolia—and began a rush that transformed the state. Oil replaced the region's agrarian Bourbons. Planters and lawyers were soon outnumbered by oilmen, refiners, and swindlers. Galveston and Houston, two of the greatest cotton ports, became global petroleum portals. Thanks to oil, Texas would send four presidents to the White House.

The next big oil rush, at Oklahoma's Glenn Pool, fertilized Tulsa, a small trading post surrounded by cattle. With money and hype, Tulsa grew by the 1930s into a bustling city of more than 100,000. The population went up and down like a yo-yo. As many as six large wells gushed every day, dropping the price of oil to 30 cents a gallon. "The greasy fluid literally flowed all over the field and costly fires were daily occurrences," wrote one journalist. "The low price of the product did not justify steel tank constructions [so] it was run into earthen reservoirs... Human life was almost as cheap as the oil that gushed from the ground." One lease purchased for $700 became a $35 million bonanza. Both the Creek and the Osage people struck it rich. Osage men drove around town in black limousines chauffeured by white fellows, while Osage women wore silk stockings and fancy leather shoes. One Native American man bought a hearse so he could sleep anywhere.

The oil boomtowns of the Southwest consisted of workers and families who followed the money. Most lived in tents, flimsy shacks called "uglies," or shotgun houses, so called because a blast through the front door would travel out the back door unimpeded. In just two years, the population of Mexia, Texas, exploded from three thousand to thirty thousand as fleets of Model Ts invaded the town. Thugs, pickpockets, thieves, prostitutes, gamblers, and fortune hunters all demanded a piece of the oil pie. Men with high-powered guns guarded "houses of infamy" with names like Chicken Shack and Winter Garden. The "Invisible Order" of the KKK responded to the mayhem by lynching lawbreakers—black or white, real or imagined— or burning them in oil. Eventually, Texas governor Pat Neff declared martial law.

Thomas Hart Benton, a Missouri artist who sought to make Americans alive to their own realities, painted *Boomtown* in Borger, Texas. He described the scene as he saw it: "Out on the open plain beyond the town a great thick column of black smoke rose as in a volcanic eruption from the earth to the middle of the sky. There was a carbon mill out there that burnt thousands of cubic feet of gas every minute, a great, wasteful, extravagant burning of resources for momentary profit. All the mighty anarchic carelessness of our country was revealed in Borger... There was a belief, written in men's faces, that all would find a share in the gifts of this mushroom town... Borger on the boom was a big party... where capital... joined hands with everybody in a great democratic dance."

A string of gushers inside the urban limits of Oklahoma City nearly destroyed the place. The wild release of gas ("like an exhaust pipe connected with hell") caused authorities to close six schools, put one-eighth of the city under martial law, and bring in two hundred state militiamen. People were forbidden to strike a match or light a fire. "In spite of these precautions," writes U.S. sociologist Rupert Bayless Vance, "the North Canadian river caught fire, burning several bridges, and 168 acres of vacant land were ignited before a thousand trained workers, wearing helmets and rubber coats, were able to cap the well."

Henry Adams, one of the greatest of U.S. historians, witnessed the acceleration of highly energized American life with some alarm in the early 1900s: "Prosperity never before imagined, power never yet wielded by man, speed never reached by anything but a meteor, had made the world irritable, nervous, querulous, unreasonable and afraid." Adams penned a famous letter to American history teachers questioning the doctrine of "indefinite progress" based on the increasing use of Daimler motors, steam engines, and coal and oil. "Man dissipates every year all the heat stored in a thousand million tons of coal which nature herself cannot now replace, and he does this only in order to convert some ten or fifteen per cent of it into mechanical energy immediately wasted on his transient and commonly purposeless objects," Adams wrote. "He startles and shocks even himself, in his rational moments, by his extravagance, as in his armies and armaments which are made avowedly for no other purpose than to dissipate or degrade energy, or annihilate it as in the destruction of life, on a scale that rivals operations of nature." Adams feared that American expansion would be followed by a contraction and the degradation of democracy. "Every gain of power... has been made at the cost of man's—and of woman's—vitality."

Sidney Armor Reeve, a celebrated U.S. steam engineer, worried about similar adjustments of the national character. Although he recognized that "the muscular system of our modern body politic is its array of energy-producing machines," he understood that energy consumption was not a "flat-footed, static thing." When engineers assembled a hundred bits into a machine, they drew on a science called mechanics. But when inanimate slaves served "a hundred million men and women... refederated into a modern State, we decline to admit that there arises therein a new and distinct form of life and energy."

But it took a global conflagration to demonstrate unmistakably the new power of oil. The United States had already commissioned the world's first oil-fueled warships, and in WWI, U.S. oil powered some of the most energy-intensive machines ever deployed: airplanes, tanks, and submarines. Petroleum was also used in the industrial production of TNT. Allied armies guzzled 650,000 gallons of oil a day. French president Georges Clemenceau opined that gasoline was "as necessary as blood in tomorrow's battles." He spoke from experience. In one famous episode, the French commandeered every truck and taxi in Paris to transport troops to the front and repel a German advance.

According to the British Admiralty, the United States supplied two-thirds of the war's oil needs. American oil firms put together a plan to ensure adequate gasoline supplies by introducing "gasolineless Sunday." American motorists sacrificed Sabbath joyriding, a popular pastime, to power the war machine in Europe. Wartime industry cooperation gave birth to a powerful new organization, the American Petroleum Institute. In the aftermath of the Great War, Henry Bérenger, the French petroleum minister, wrote a 1921 memorandum that laid out the transformative power of petroleum: "He who owns the oil will own the world, for he will rule the sea by means of the heavy oils, the air by means of the ultra refined oils, and the land by means of petrol and the illuminating oils. And in addition to these he will rule his fellow men in an economic sense, by reason of the fantastic wealth he will derive from oil—the wonderful substance which is more sought after and more precious today than gold itself." President Coolidge would later proclaim that "the supremacy of nations may be determined by the possession of available petroleum and its products."

BY 1920, oil had saturated every aspect of American life and the United States mistook its geological luck for global exceptionality. In addition to powering locomotives, ships, cars, and airplanes, oil was used to make more than 250 products, including paints, waxes, soaps, and artificial butter. *The Evolution of Oil,* a typical polemic of the day, saluted oil as a divine force that aided manufacturing, multiplied comforts, and standardized warfare. For centuries the "devil's tar" had remained an unutilized curiosity, until "American ingenuity and adaptability... made it the marvelous agent that it is today." In 1921, U.S. engineer Joseph E. Pogue, in *The Economics of Petroleum,* reflected on the speed of change with some caveats. Only cotton and breadstuffs still outvalued U.S. petroleum exports to Chile, Italy, the United Kingdom, Canada, and France. A million barrels now pulsed through pipelines from fields in Texas, Oklahoma, California, and Louisiana, to five hundred refineries. Automotive transportation had grown at an astounding rate of 40 percent per year for a decade with "no industrial parallel." Cars accounted for 25 percent of the volume and nearly 50 percent of oil's value. But low prices, waste, and carelessness seem to characterize "American economic practice," Pogue wrote. Industry routinely left more than 60 percent of the oil in the ground and stupidly burned off natural gas. The exhaust pipe of the average automobile contained nearly 30 percent of the heat from the gasoline. Pogue reckoned that 40 million of the 160 million barrels of fuel oil burned in the United States in 1917 could have been saved through more intelligent operation. He worried that "American petroleum has been brought into service at a tremendous cost of the oil itself."

Geologists, for their part, worried about the depletion of oil fields. Most "rock hounds" reckoned that the United States had only a twenty-year supply left in the ground. They urged the U.S. government to encourage oil companies "to acquire foreign sources of supply wherever available." Without government backing on this effort, some geologists worried, "the future is fraught with hazard to an industry that stands as a monument to American organizing genius." And so the United States sent out its oil pioneers to conquer the world's petroleum frontiers. Engineers and roustabouts peopled the industrial crusade, with the odd East Coast banker or accountant in tow. Everywhere the Americans went, they mechanized the landscape and made everything beholden to black gold. Edward Doheny opened up Mexico's veins. Standard Oil invaded Venezuela. "The American oil driller is a distinct type," wrote journalist Isaac Marcosson in 1924. "You can spot him anywhere, whether he is knee-deep in Galician mud or emerging from the mist of the Slavic steppes. He has set up a little Texas, Oklahoma, or California wherever he has gone, no matter how remote. His courage and character have been a credit to his country." The spirit of this oil expansion embodied the 1907 vision of Elihu Root, secretary of state under Theodore Roosevelt: "Our surplus energy is beginning to look beyond our own borders, throughout the world, to find opportunity for the profitable use of our surplus capital, foreign markets for our manufactures, foreign mines to be developed, foreign bridges and railroads and public works to be built, foreign rivers to be turned into electric power and light," Root wrote. Oil would allow the United States to conquer the world the same way the nation had settled the American West.

On home soil, oil had begun its remarkable conversion of California. At the turn of the century, over two hundred companies drilled inside the city limits of Los Angeles, the world's first oil port. During the 1920s, the state produced nearly a quarter of the world's oil. California's population grew by 365 percent between 1900 and 1940, but the state's five oil-field counties expanded even faster, clocking in at 1,200 percent. The boom gave the region the boost it needed to eventually become one of the globe's top ten economies.

Oil also did for California what it later did for Alaska: made the state an advertisement for easy living. Decades later, journalist Eric Schlosser would call Southern California "the Kuwait of the Jazz Age." Petrodollars built roads to encourage more cars and funded Hollywood movies. The flow of cash corrupted politicians and underwrote extravagant estates. In the first of many sensational political scandals, California oilmen Edward Doheny, of Pan American Petroleum, and Harry Sinclair, of Mammoth Oil, "loaned" U.S. secretary of the interior Albert Fall $100,000 in 1922 to gain access to lucrative oil leases in Elk Hills, California, and Teapot Dome, Wyoming. The U.S. government had previously set aside the reserves for its navy. Reporting on the story, *Time* magazine declared that oil "lubricated the jaws of the nation. Newspapers screamed it, preachers damned it, Mr. Average Citizen swallowed it and was shocked."

Southern California provided the template for the suburban sprawl that would later redesign much of the United States. In some parts of the state, where one in four people owned a car, gas stations dotted the highways every 2.6 miles. California also showcased the industry's ability to promote more oil spending with user-financed road building; highway taxes became so profitable that the government virtually abandoned the idea of public transit. Until the 1940s both San Francisco and Los Angeles possessed well-used electric trolley systems. But Big Oil changed that. A shell firm financed by General Motors, Exxon, and Firestone Tire bought San Francisco's trolley system and then sold it with draconian conditions attached: the trolley cars had to be replaced with General Motors buses running on Exxon gasoline and Goodyear tires. General Motors, Standard Oil of California, and Firestone similarly dismantled L.A.'s electric train service. As journalist Carl Solberg would explain in *Oil Power* in 1976, "In this fashion the big companies blotted out 100 electric railway systems in 45 cities—including New York, Philadelphia, Baltimore, St. Louis and Salt Lake City. In 1949, they were convicted of criminal conspiracy. The ring leader, General Motor's treasurer, H. C. Grossman, was fined one dollar." To placate opponents of coastal drilling (and ultimately oil companies that wanted to drill horizontally into oceanfronts), California used oil royalties to create public beaches and redwood parks. Only automobiles, the main market for oil, made a visit to these places possible.

It was in California, too, that oil met and embraced a new form of religious extremism. Lyman Stewart, the president of Union Oil and a devout Presbyterian, poured his fortune into spreading the word of the Gospel as unerring truth. With his brother Milton, Stewart anonymously funded and distributed more than three million copies of *The Fundamentals,* a collection of essays that defended the Christian faith by attacking evolution, liberalism, progressivism, socialism, and Darwinism. Disturbed by oil workers' profanity, Stewart built chapels close to oil derricks; one of his fields was called Christian Hill. His under-the-radar activism would later produce a fundamentalist president: oilman George W. Bush.

Critics of the U.S. oil boom emerged early on, though their voices were usually drowned out by the industry's relentless boosterism. In his 1926 classic, *The United States Oil Policy*, Kansas economist John Ise reckoned that a *Scientific American* writer got it right when he compared America's profligate oil spending and relentless draining of domestic oil fields to "the flight of a swarm of locusts across a fruitful land." Ise, an old-fashioned conservative, also liked the story Gifford Pinchot told about a man stranded with one barrel of water on a boat that had drifted twenty days from shore: "What should we think of that kind of a man, if under these circumstances, he not only drank all the water he wanted, but used the rest of it to wash his hands?" That was the American oil consumer, wrote Ise. To Ise the automobile symbolized unprecedented extravagance. The joyride, he wrote, was not a "pursuit of happiness" but a stealing of energy from future generations. "We probably have a right to prefer our thousandth joy ride to the thousandth joy ride of our grandchildren, but whether we have the right to deprive them of their only ride in order that we may indulge ourselves with two thousand such rides is another question." In a summation that could have been written today, Ise concluded that "the history of oil exploitation in the United States is a history of criminally rapid, selfish, and wasteful use of an exhaustible resource which, as far as present knowledge goes, will be indispensable in the economic lives of the next generation, as in our own."

Two years earlier, in 1924, American journalist Isaac F. Marcosson had published *The Black Golconda*, an account of his visits to various U.S. oil fields and the global oil patch. Marcosson's book provides a lively snapshot of the oil industry, which he called "a supreme necessity." His frankness and clarity might startle even modern-day gasoline consumers. For starters, he worried about the United States' ability to maintain its supremacy in the trade. As American, Dutch, and British oil companies battled for new global real estate, Marcosson realized, the prize lubricant had become an international irritant: "What scientists call the petroliferous epoch is in reality a pestiferous period because of the grand scramble for that product of Nature which has been well called flowing gold." American car owners consumed over 5 barrels of oil a year each compared with .18 for the rest of the world, Marcosson wrote, yet they knew nothing about the industry supplying their billion-dollar-a-year gasoline addiction. The average John Jones was totally ignorant of "the hazards and handicaps that beset the evolution of crude petroleum after its release from long imprisonment in the bosom of Mother Earth until it gushes forth to heat, light and propel this humming universe."

Like John Ise, Marcosson could see the economics of depletion at work. By 1924, the U.S. boasted 300,000 producing wells and was drilling 67 new ones a day. Shallow wells that cost mere thousands of dollars to drill had been replaced by 2,000-foot wells that cost $300,000 each. One-fifth of them came up dry. "Just as we pioneered the world," Marcosson wrote, "so have we supplied the universe. The principal drain, however, has come from our own phenomenal demands, born of American temperament, initiative, and expansion. Though many other countries—and they include China, Japan, Burma, Russia, Galicia, and Rumania—were ahead of us historically in the commercial utilization of petroleum, we—and it is typical of our instinct for waste—have gorged ourselves with the product, with the inevitable result."

When British social critic G.K. Chesterton visited this new material kingdom in the 1920s, he was impressed by its brashness. But he was also stunned by people's zealous fondness for "selfish sensational self-advertisement." The problem was this, he said: "A number of people who were meant to be heroic and fighting farmers, at once peasants and pioneers, have been swept away by a particular pestilence of a particular fad or false doctrine; the ideal of which has and deserves the detestable name of Making Good." Americans didn't want to venerate "the Oil King" Rockefeller, wrote Chesterton; "they wanted to imitate him."

So did economists. Irving Fisher, a Yale professor and the son of a preacher, defined economics as "the science of wealth" and thought it should become the United States' new physics. During the Roaring Twenties, Fisher was one of America's most quoted economists. Unlike Veblen, who thought engineers should run the economy because they understood energy, Fisher championed money itself as the road to prosperity, preaching that mathematics was the lantern that would dispel "old phantasmagoria" and make the future predictable. Like the slaveholders of old, Fisher also advocated for the sterilization or elimination of people he called "life-waste" from the economy so that the "fit" could prevail. In 1929, he put all of his money into the stock market, declaring that Wall Street had struck a "permanently high plateau."

Beginning in 1930, the U.S. oil industry too nearly died in a gusher of waste, theft, and overproduction. After the discovery of the East Texas oil fields, independent drillers invaded Kilgore, Texas, like some hungry horde. In just three years, the number of derricks exploded from 3,540 to 12,000.

Eager for quick profits, the boomers flooded U.S. markets with a million barrels (more than half of U.S. demand), driving prices down from $1.10 to 10 cents a barrel in the middle of the Depression. Water cost more than oil in East Texas, and filling stations typically offered a dozen eggs or a free chicken dinner with a fill-up. The glut threatened the competitiveness of the country's higher-cost oil fields and stripper wells, which produced barely 10 barrels a day. When Texas tried to control production, the independents cried foul and smuggled their oil out of state. As much as 100,000 barrels of illegal or "hot" oil left Texas every day. To restore order, the governors of Oklahoma and Texas declared martial law and sent in thousands of National Guardsmen. The crisis grew so protracted that U.S. interior secretary Harold Ickes proposed to President Roosevelt that he set up an "oil dictatorship" to investigate the industry, limit production, and set minimum prices.

After lengthy legal and political battles, the United States established the Interstate Oil Compact in 1935, ending the roller coaster of surplus and glut by stabilizing prices for decades. The shape of the oil-price graph changed from a yo-yo to an orderly ascending staircase. The major companies agreed to limit production in the name of conservation. The federal government banned hot oil and produced reliable consumption statistics to help establish state quotas. To keep peace in rural areas, rationing policies generally excluded the low-producing wells, most of which were owned by independent producers. Texas, the largest oil-producing region in the world, oversaw its price controls mostly through its Railroad Commission. The unusual system, which most Americans still know nothing about, directly subsidized high-cost marginal domestic producers but guaranteed that U.S. oil demand would be met by U.S. production. (It basically ended in the 1970s.) It was a "drain America first" policy that restored prices to a dollar a barrel and would provide consumers with stable oil prices until the 1970s yet guarantee oil producers higher-than-free-market prices. The system proved so successful that it served as the model for OPEC.

In 1933, the first American wildcatters landed in Saudi Arabia. "The prize," in unbelievable quantities, was found by Standard Oil of California and Texas Oil's CALTEX (it would become the Arabian American Oil Company, ARAMCO), created to mine the desert. "Within less than 10 years, Saudi Arabia had learned to be dependent to an almost alarming extent upon the income from oil; already the nation's wants had proliferated under American example," wrote the novelist Wallace Stegner in his little-known 1956 history, *Discovery!*. Stegner compared ARAMCO's social engineering of Saudi society to the impact of Europe's Marshall Plan. A couple of palm-thatched fishermen's *barastis* in al-Khobar became "a solid half mile of glass-fronted, electric-eyed, mercury-vapor-lighted shops, bottling works, cement plants, garages and what not." Thanks to oil, ARAMCO housewives could buy French perfumes, Danish furniture, and tranquilizers in a desert.

The importance of the world's largest oil reserves after Texas did not escape the attention of President Franklin D. Roosevelt. In 1943, he recognized Saudi Arabia as a national interest. Though ill and spent by the war effort, Roosevelt devoted two entire days to meeting with Ibn Saud, the nation's ruler, aboard a U.S. battleship. The leaders impressed each other. Roosevelt gave Ibn Saud a plane and a wheelchair. The two petro-kingdoms have played a game of master and slave ever since.

Oil and its deadly inanimate helpers dominated the Second World War just as they had the First. While Americans were busy taking their petroleum abundance for granted, Japan and Germany, as oil importers, had studied synthetic fuels and stockpiled U.S. supplies. (Hitler had a high-energy vision for Germany that included "people's cars" and highways that would "usher in a new traffic epoch, the time of the automobile.") The two expanding nations both recognized that a shortage of petroleum would limit their economic and political ambitions. (In 1938, Germany consumed 44 million barrels of oil annually, compared to the United States' 1 billion.) And so both gambled on unique military strategies to obtain foreign energy supplies. German U-boats sank oil tankers in the Atlantic, and after the United States embargoed Japan's oil, the Japanese bombed Pearl Harbor. During the conflict, to compensate for their petroleum poverty, both Japan and Germany employed human slavery on a massive scale. Japan enslaved millions of Chinese workers, and the Nazis shackled millions in Eastern Europe and the Soviet Union. The United States concentrated its own oil spending on military hardware.

Germany, with few domestic oil sources, depended largely on coal liquefaction factories. The Nazis actually based their key military strategy, the blitzkrieg, on Germany's lack of cheap oil. Quick strikes in Norway, France, and Poland captured more oil supplies. But the strategy ultimately failed when the Germans could not quickly secure Russia's two largest oil fields.

German planes, which ran on poor-quality synthetic fuels made from coal, lacked the maneuverability and speed of Allied machinery running on high-octane fuels. In the end, the contest often boiled down to how many mechanical fighting slaves the combatants could afford to mobilize. Submarine warfare, a high-energy petroleum adventure, tells the story. Each side tried to use its submarine fleet to cut off oil supplies. To combat the German U-boat menace, a meager yet deadly

$2 billion enterprise, the Allies eventually spent $26 billion in iron, oil, and ships. Meanwhile, the submarine war in the Pacific cost the Japanese $1 trillion. After the war, Japan's munitions minister, Toyoda Soemu, admitted that "the shipping shortage and the scarcity of oil were the two main factors that assumed utmost importance in Japan's war efforts." In 1946, the U.S. military confidentially interviewed some of Germany's highest-level officers and asked them why they had lost. To a man, they cited oil quality or shortages.

Prior to the war, Japan had imported nearly 80 percent of its oil from the United States. The imperial wannabe attacked Pearl Harbor with aviation fuel purchased in California. But despite a two-year stockpile and the capture of Dutch-owned oil fields, the Japanese navy and air force soon ran out of fuel. Desperate, they turned to substitutes made from potatoes, peanuts, coconuts, castor beans, and pine roots. (It would take nearly a million root-pickers to produce 12,000 barrels a day of pine oil; Japan never got beyond 75,000 barrels a month.) The Japanese even developed kamikaze, or suicide, air attacks to conserve as much oil as possible.

The war cemented ties between the U.S. government and the country's powerful petroleum industry. President Roosevelt divided the country into five Petroleum Administration Districts and grabbed seventy-two executives from the oil industry to serve on the Petroleum Industry War Council. One of the council's most important tasks was the construction in 1943 of two pipelines, Big Inch and Little Big Inch, from Texas to the east coast. Without them, the Allies would never have had enough oil to invade Europe. Texas pipeline builders followed General Patton and his tanks every inch of the way.

Meanwhile, faced with shortages of coal, salt, and wood—the usual ingredients for any chemical product—industry played with petroleum. U.S. chemists dabbled with oil to find new properties and discovered magical uses. They combined styrene with butadiene to make synthetic rubber, and thereby ended rubber shortages. They synthesized TNT (trinitrotoluene), referred to by *Popular Mechanics* as "one of the most powerful but obedient of all war explosives"; making toluene from petroleum instead of coal tar allowed the U.S. to produce more warheads. Next came a variety of plastics, fertilizers, and pesticides. Eventually the petrochemical industry would turn out as many as four thousand different products, including garbage bags, balloons, credit cards, fishing rods, and waterproof clothing. Chemical detergents dethroned animal-based soaps, and polyester challenged cotton. Anything that could be made with cheap oil took over the marketplace.

Near the end of the bloody contest (60 million dead), U.S. anthropologist Leslie White soberly reflected that a B-17 (Flying Fortress) bomber flying over Europe consumed more energy during just one flight than the continent's entire population had during the Stone Age. White concluded that fossil fuels and their technological forces had created a unique "Power Age." Cultures advanced when they tapped into an energy source that allowed them to "burst asunder the social system which [bound] it," he wrote. But he did not see the United States' future as preordained: "The triumph of technology and the continued evolution and progress of culture are not assured merely because we wish it or because it would be better thus... Culture will decline unless man is able to maintain the amount of energy harnessed per capita per year by tapping new sources."

The United States rebuilt the war-ravaged economies and infrastructure of Europe and Japan largely with oil. What oil-powered machines had turned into rubble, petroleum-based industries could now raise from the dead. Oil also broke the back of powerful coal unions—the ambitious Marshall Plan, a $13 billion program funded by U.S. taxpayers, stipulated that Europe and Japan must replace coal with oil as their key energizers. The money built roads, purchased U.S. cars, and subsidized European carmakers too. One-sixth of the plan's funds directly paid for oil pumped by U.S. oil companies operating in the Middle East. (According to Carl Solberg, firms like Exxon, over a fifteen-year period, took more wealth "out of the Middle East than the British took out of their empire in the entire nineteenth century.") As laid out in the Marshall Plan, U.S. multinationals sold the world's cheapest oil at prices 40 percent higher than that of Texas crude. And the restoration project worked another one of petroleum's economic miracles: Japan and Europe experienced economic renaissances. Europe's GNP rose by 32.5 percent after 1947, and industrial output increased 40 percent beyond prewar levels. Agriculture boosted production by 11 percent. The Europeans called it "supergrowth." By 1960, oil's share of Europe's energy diet had risen from 10 to 30 percent.

While the Marshall Plan petrolized Europe, Norman Borlaug took oil and industrialized global agriculture. Recruited to go to nationalistic, petroleum-exporting Mexico in 1944 as part of a joint venture to research wheat production, Borlaug changed the energy of plant growth: the Iowa-born plant breeder shrank grains so that the plants directed more energy to making heads full of seedy protein. The new high-yielders grew so fast that they needed more water and more soil nutrients. Borlaug's so-called Green Revolution depended on oil-based fertilizers, pesticides, and diesel-run irrigation systems. Without petroleum, the new agricultural system was a car without an engine. Aided by the Ford and Rockefeller foundations, Borlaug developed other high-yielding grains in India, Pakistan, and China. Between 1950 and 2000, his industrial techniques increased agricultural productivity sevenfold. "I want science to serve a useful purpose," said Borlaug, who went on to win a Nobel Prize. Today, his high-yielding grains feed billions.

Petroleum continued its conquest of the United States. President Dwight Eisenhower, impressed by Hitler's idea of a national superhighway system, began the construction of a $27 billion interstate highway network in the 1950s. It was intended partly as a defense system in case U.S. cities had to be evacuated during a nuclear attack. Finally completed in 1991, the 47,000-mile highway system became "the largest engineering project the world has ever known." Trucks replaced railroads as the movers of goods, and urban freeways criss-crossed cities. In 1955, before construction began, the average American consumed 202 gallons of gasoline to drive nearly 3,000 miles. By 1975, drivers were eating up 343 gallons to go an average of 4,481 miles. American consumers, who collectively traveled 1 trillion miles a year, now worked to drive.

As highway construction got underway, the United States invented an entirely novel form of urban living: suburbia. One of the first suburbs, predating the superhighways but anticipating the development they encouraged, was Levittown on Long Island. Built on potato fields, the standardized community, planned for seventeen thousand residents, provided cheap, identical three-bedroom boxes that could be erected in twenty-seven easy steps. Each came with an oil-burning furnace, a washer and dryer, and a garage. Each of the development's ten neighborhoods boasted an elementary school, a pool, and a playground. Returning veterans flocked to Levittown, which became known as "Fertility Valley" and "the

Rabbit Hutch." Business writer William Whyte called suburbia "the dormitory of the new generation of organization men." Herbert Gans, a sociologist who lived in Levittown for two years as a "participant observer," praised the new form of living. Although "the critics have argued that long commutation by the father is helping to create a suburban matriarchy with deleterious effects on the children," wrote Gans, "and that homogeneity, social hyperactivity, and the absence of urban stimuli create depression, boredom, loneliness and ultimately, mental illness," suburban life actually produced "more family cohesion." Gans dismissed the disappearance of farmland near big cities as "irrelevant now that food is produced on huge industrialized farms, and the destruction of raw land and private upper class golf courses seems a small price to pay for extending the benefits of suburban life to more people." By the 1960s, these car-dependent "packaged villages," as Whyte called them, had pulled more than 30 million Americans from major U.S. cities. It remains one of the greatest lateral migrations in North American history.

In 1957, Admiral Hyman Rickover, father of the nuclear submarine, gave a sobering speech to a group of U.S. physicians in St. Paul, Minnesota. Rickover, who had served the republic since 1912, keenly appreciated how oil had reshaped the national character, and he encouraged his audience to consider soberly their responsibilities to "our descendants—those who will ring out the Fossil Fuel Age." Responsible living, he said, meant energy conservation, excellent education for all citizens, a new culture of self-denial, and higher taxes to fund a larger, more complex United States. The alternative was doubt, indecision, chaos, and collapse.

Rickover began with a simple story. Only a century earlier, fossil fuels provided but 5 percent of the world's energy;

humans and animals did most of the rest of the work. By 1957, fossil fuels supplied about 93 percent of the world's energy. With only 6 percent of the world's population, the United States consumed nearly a third of its fossil fuels.

The admiral reminded his audience of the power that most Americans took for granted: "The enormous fossil energy which we in this country control feeds machines which make each of us master of an army of mechanical slaves," he said. Machines furnished every American industrial worker with energy equivalent to that of 244 men, while the equivalent of at least 2,000 men pushed that worker's automobile along the road and another 33 served as faithful household helpers. Each locomotive engineer controlled energy equivalent to that of 100,000 men, each jet pilot that of 700,000. "Truly the humblest American enjoys the services of more slaves than were once owned by the richest nobles," said Rickover, "and lives better than most ancient kings. In retrospect, and despite wars, revolutions, and disasters, the hundred years just gone by may well seem like a Golden Age."

Rickover told his listeners that he didn't know of a society where a reduction in energy slaves—human or petroleum—had not resulted in a decline of civilization. "Our civilization rests upon a technological base which requires enormous quantities of fossil fuels," he continued. "What assurance do we then have that our energy needs will continue to be supplied by fossil fuels? The answer is—in the long run—none."

Rickover reflected that the United States had begun in 1776 as a nation of 4 million people and boundless resources. "We conserved what was scarce—human labor—and squandered what seemed abundant—natural resources—and we are still doing the same today. Much of the wilderness which nurtured what is most dynamic in the American character has now been buried under cities, factories and suburban developments where each picture window looks out on nothing more inspiring than the neighbor's back yard with the smoke of his fire in the wire basket clearly visible." The concentrated size of cities, government, and corporations now demanded "an ever larger share of what we earn to solve problems caused by crowded living." The United States must undergo a transition, Rickover concluded. "Fossil fuels resemble capital in the bank. A prudent and responsible parent will use his capital sparingly in order to pass on to his children as much as possible of his inheritance. A selfish and irresponsible parent will squander it in riotous living and care not one whit how his offspring will fare."

Three years later, a departing President Eisenhower issued his country another kind of warning. The United States, he said, now had a vast defense establishment that employed more than 3 million people and spent more money annually than all the corporations of the country combined. This "military-industrial complex," a creation of surplus petroleum energy, influenced the economic, political, and spiritual life of every American, said Eisenhower, and "we must not fail to comprehend its grave implications." He described the "potential for the disastrous rise of misplaced power" as a real threat. But few took Eisenhower's vision seriously. Today, at 400,000 barrels a day, the U.S. military remains the largest single institutional consumer of oil on the planet. U.S. troops in Iraq used four times more fuel during the Iraq War and subsequent occupation than had General Patton's Third Army in Germany.

Petroleum changed the vocabulary and character of the whole of the United States, much the same way slavery had for the U.S. South. In 1962, Yale historian George Pierson declared that the United States owed its superiority to what he called the M-Factor: "movement, migration and mobility." People were comers or go-getters or goners. The phrases that mattered said it all: "Get going!" "Don't be a stick-in-the-mud." "I don't know where I'm going, but I'm on my way."

But only ninety-two miles from the scene of the first oil boom in Titusville, Pennsylvania, resistance was blossoming. It began with Rachel Carson, a fisheries biologist from Springdale. In 1962, in *Silent Spring,* Carson exposed how the $300 million oil-spawned pesticide industry was deforming young birds and fish. E.B. White called her critique "an Uncle Tom's Cabin of a book,—the sort that will help turn the tide." Not surprisingly, the oil industry and its chemical associates branded the ecologist a Communist and a "fanatic defender of the cult of the balance of nature." They spent millions trying to discredit and defame her. But Carson's book, combined with a 1969 Santa Barbara oil spill, launched the environmental movement. Wrote Carson later in life, "The question is whether any civilization can wage relentless war on life without destroying itself, and without losing the right to be called civilized."

AS MANY had predicted, the U.S. oil boom was already reaching its limits. The American petroleum and economic machine quietly peaked in the 1970s, though the evidence is not yet all in. U.S. domestic oil production reached a height in 1970 that has not since been replicated. Then it settled into persistent decline. Between 1970 and 1974, U.S. dependence on foreign oil grew from 21 percent to 36 percent. Over the next couple of decades, the country would spend trillions of dollars to buy oil. Petroleum economist James Hamilton notes that "the real price of oil rose 8-fold from 1970 to 2010, while U.S.

production of oil fell by 43% over those same 40 years." The United States now imports about 50 percent of its oil, at a cost of nearly a billion dollars a day.

The world's former petroleum master has become another struggling consumer dependent on either foreign oil or costly unconventionals. North American sources include tough oil pulled from two miles below the ocean floor, high-cost North Dakota shale oil, and dirty Canadian crude from the Alberta tar sands with the consistency of asphalt. All of these take more capital, more energy, and more carbon making to produce than conventional oil. Due to oil-price volatility, car sales and car mileage per capita are declining rapidly. A housing bubble built on cheap oil has burst, and Americans now forage for fruits and nuts in the backyards of foreclosed McMansions. The aviation industry, its infrastructure declining, grapples with high fuel prices and declining sales. The public highway infrastructure that oil built rots and decays; in 2009, the American Society of Civil Engineers gave it a D–. The U.S. military, the world's largest consumer of petroleum, has established new programs in an attempt to green its fuel train. The world's most powerful industry, with its motto of "Drill, baby, drill," continues to battle any transition with the vehemence of a Southern slaveholder. And the world's wealthiest citizens are busy attending to their petroleum slaves.

4

The New Servitude

. . . . .

"We think about machines, we talk about them, but neither thinking nor talking can help us."

JOHANN WOLFGANG VON GOETHE,
Wilhelm Meister's Journeyman Years, 1821

IN HIS BOOK *The Mansions and the Shanties*, the great Brazilian historian Gilberto Freyre wrote floridly about three centuries of slavery, patriarchal families, and single-crop economies. It took a determined British invasion of "steam-powered energy" to liberate blacks and begin a gradual redistribution of power. But what amazed Freyre was how long slavery persisted in the presence of "steam horses." Moral arguments against the institution stalled in the face of the energy habits "demanded by old civilizations." Talk of replacing slaves with machines or animals made the patriarchs "dizzy." While nineteenth-century Europeans and North Americans rode in carriages, Brazilian planters still championed palanquins carried by trotting slaves. Wagons and oxcarts did not busy the roads because there were no roads. Brazil's energy infrastructure supported only trails, upon which "Negros" carried bags of cotton or sugar on their heads. The masters even rented out their slaves, as though they were mules.

Brazil's slaveholders showed "a lack, or near lack, of feelings of pity" for abused slaves, Freyre wrote. A slaveholding patriarch might harbor some degree of human sentiment for black nurses and household help, but toward "beasts of burden" the master showed what one visiting German prince described as a "complete lack of the idea or sentiment of conscience." The easy living afforded by slavery explained this lack of conscience, just as the easy living afforded by petroleum might explain North American indifference today to the proliferation of inanimate slaves in our midst.

Noisy leaf-blowers, expensive SUVs, and glowing smartphones dominate modern life as fully as did the servants in a nineteenth-century Brazilian "Big House." The average North American or European consumer thinks of these inanimate servants as entitlements. And although our comfort providers and labor savers number in the billions, we largely pretend that they do not exist. U.S. plantation owners at least earnestly debated the morality of living off the sweat of their shackled servants. Their modern descendants take immediate offense at any discussion about the carbon emissions of their mechanical servants.

In 2009, one British family living in a four-bedroom house became the subject of a subversive energy experiment. While the unsuspecting foursome flicked on gadgets one Sunday with the abandon of Roman patricians, an army of volunteers, the Human Power Station, furiously pedaled one hundred bicycles next door to generate the family's needed energy. At the end of the day, the jaws of these modern slaveholders dropped when a BBC television crew introduced them to the exhausted slaves who had boiled their tea. It had taken twenty-four cyclists to heat the oven and eleven to make two slices of toast. Many of the cyclists collapsed. Several couldn't walk for days. Not only that—the pedalers actually consumed more energy, in food, than they generated.

The experiment crudely illustrated widespread consumer ignorance about energy spending. It also convinced one of the experiment's designers, Tim Siddall of Electric Pedals, that "volunteer slavery" (hordes of sweating cyclists) or old-fashioned shackled labor might be needed to power the future: "I have no doubt that slavery will return as the world's energy resources get increasingly scarce," the *Guardian* quoted him saying.

Most people don't regard oil as fuel for energy slaves, but that's what the master resource has become. Thanks to the work now performed by energy slaves, North Americans behave, think, and often look like obese, overbearing plantation slave owners. Energy slaves, of course, are more portable and versatile than human muscle. They not only grow and deliver food, but also transport people and goods and energize fields and cities. They also consume energy. The automobile, the world's favorite mechanical slave, consumes nearly a quarter of U.S. oil supplies every year. Every laptop computer, a veritable brain slave, arrives impregnated with the 240 kilograms of oil needed to make it. Like any servile companion, oil removes the toil. But just when the era of cheap fuel for inanimate slaves has ended, the complexity of social and political arrangements for these slaves has reached new heights.

Roman masters counted to the gram how many calories were needed to keep their slaves running. The average North American has no idea how many barrels, joules, or watts of energy he or she must spend to keep the servants humming.

But the numbers are sobering. David Hughes, perhaps Canada's premier energy analyst, calculated in 2011 that one barrel of oil contains approximately 6 gigajoules (6 billion joules), or about 1,700 kilowatts, of energy. A healthy individual on a bicycle or treadmill can pump out enough juice to light a 100-watt bulb, about 360,000 joules an hour. Accounting for weekends and holidays off and a sensible 8-hour workday, Hughes figures that it would take one person 7.37 years on a bicycle or treadmill to produce the dense, highly portable energy now stored in one light barrel of oil. If the person ran or rode 12 hours a day, 7 days a week, with no holidays, says Hughes, a barrel of oil would be equivalent to 3.8 years of human labor. Given that the average North American now consumes 23.6 barrels of oil a year, every citizen employs about 89 virtual slaves. A family of 5 commands nearly 500 slaves. A nation of 300 million controls an incredible phalanx of 27 billion largely mechanical and oil-fed workers. The Italian anthropologist Alberto Angela dryly observes in *A Day in the Life of Ancient Rome* that a cup of gasoline is equivalent to the energy of 50 slaves pulling a Fiat for 2 hours.

Modern inequities abound in this petroleum slavocracy. Most of the world has few or no slaves; North Americans and Europeans hoard the majority by burning the most oil. The average Canadian "consumes 5 times the world average per capita consumption, 7 times the per capita consumption in China and 29 times the per capita consumption in India," said Hughes in an interview. The new and leaner slave masters of Shanghai or Tianjin burn 2.4 barrels of oil per capita a year, which puts 9 coolies at their beck and call. Hughes has no doubt that "we have been less cognizant of the services provided by fossil fuels than people were of those from their slaves... Slavery, after all, was in your face. Now it's all about filling up the tank."

To the great energy historian Vaclav Smil, the image of hundreds of slaves toiling in our homes and workplaces does not fully capture the realities of a high-energy culture. Even a household of Roman slaves, he says, "could not provide energy services comparable in terms of convenience, versatility, flexibility with those delivered by electricity and fossil fuels: no filling of oil lamps, no fiddling with wicks, no preparation of kindling. No lighting of charcoal, no frequent stoking of a wood fire, no heavy smoke from insufficiently air-dried wood, no scraping out of ashes for a bread oven, no carrying of heavy shoulder loads, no dragging of a recalcitrant mule or donkey—just a flip of a switch, a turn of the key, numbers punched into the thermostat."

Smil recently compared the petroleum-assisted world of a U.S. farmer with that of a Roman counterpart. In less than two hours of labor, a U.S. Midwest grain farmer can produce a ton of wheat, all while sitting in a giant air-conditioned vehicle. The most resourceful Roman farmer needed 350 hours of labor to produce a ton of grain over a year. A modern American farmer assisted by inanimate slaves could provide, in a day's work, enough bread to feed 180 Romans for a month. In contrast, a Roman slave or peasant could "barely supply a monthly ration for a single household slave" over the course of a year.

The ubiquity of energy slaves in the United States today would surely astound Harriet Beecher Stowe, the author of *Uncle Tom's Cabin*. The United States traded in an energy system that shackled Topsy, Sambo, and Eliza for one that enslaves consumers to Hummers, Lincolns, and Fords. (Tellingly, more than 40 percent of product names for automobiles salute slavery's legacy, including Stylemaster, Regent, Royal, Patrician, Master De Luxe, and Powermaster.) Americans now feed 50 million lawnmowers, 2 million snowmobiles,

6 million motorcycles, 13 million boats, 8,000 commercial jets, and 250 million cars, vans, and trucks—and China, India, and Brazil can't wait to own their mechanical complement. Most of these energy slaves are judged by their weight, looks, and strength. An extraterrestrial visitor might conclude, writes Vaclav Smil, "that machines are the planet's dominant creatures, with humans as their servants."

The car illustrates the new servitude grandly. Although the oil industry argues that "the automobile is the most practical and democratic transportation device in history," the facts tell a more nuanced story. In 1974, Ivan Illich, a Catholic theologian and social critic, examined the full cost of car ownership, including the hours needed to buy, operate, and repair the vehicle. He added up insurance payments, hospital visits, traffic tickets, road congestion, and the like. On average, Illich figured, a typical American male devoted 1,600 hours a year to his car to drive only 7,500 miles. That worked out to 4.7 miles an hour, the speed of a very brisk walk. (Tests in many large cities have found that the average speed of automobiles in congested areas is actually lower than that of a horse and buggy.)

Smil, revisiting Illich's numbers, found that Americans now spend even more on their oversized vehicles (a typical Hummer, a warrior slave no longer in production, weighs 4,700 pounds) to go only 3 miles an hour. Even a retired person can walk faster than that. "We know that anorexia nervosa correlates highly with high incomes," wrote the energy expert in a splendid 2006 essay, "and so in affluent neighborhoods of U.S. cities we can see nearly 5,000-kg cars driven by anorexic 50-kg females to buy a 500 g carton of a slimming concoction."

Road congestion, car accidents, and exhaust pollution subtract $30 billion U.S. from the British economy every year.

Emissions from energy slaves in the United States totaled 346 million tons of greenhouse gas emissions in 2004, enough carbon to fill a coal train 55,000 miles long. Every day, 255 million U.S. drivers eliminate an average of 100 cyclists and walkers from the streets. (Pickups and SUVs slaughter the most people and come with the highest crash costs.) Globally, car accidents cripple 23 million people every year; bury a million, mostly young folks; and cause some $600 billion worth of damage. Asia is the most lethal region for walkers and cyclists.

The ascendance and transformative powers of these inanimate slaves caught the attention of the engineer, architect, and futurist Buckminster Fuller decades ago. In the early 1940s, it was Fuller who coined the term *energy slave*. After combing through energy statistics compiled by the U.S. and British armies, he arrived at some stunning conclusions. In 1810, the population of the United States included one million families and one million slaves. Even though few Americans owned slaves, that worked out to one slave per family. By 1940, coal- and petroleum-fired mechanical energy had placed approximately thirty-nine energy slaves at the service of every American citizen. Moreover, said Fuller, these new slaves could "work under conditions intolerable to man, e.g., 5000° F, no sleep, ten-thousandths of an inch tolerance, one million times magnification, 400000 pounds per square inch pressure, 186000 miles per second alacrity and so forth."

Fuller, a charismatic fellow, knew that he had stumbled upon something dramatic: "Suddenly you have two hundred non-human slaves doing the work. An enormous step up in the standard of living is represented, as well as doing away with the inhumane idea of the human being the muscle machine to be commanded." To illustrate the social and political transformation, he published his World Energy Map in *Fortune* magazine in 1940. The map showed where the world's

2 billion people lived with white dots. Red dots illustrated hubs for energy slaves. "Mechanization, the harnessing of energy, is man's answer to slavery," reported Fuller. Based on total energy consumed from mineral sources and waterpower and the energy output of one human per year, Fuller estimated that approximately 36,850,000,000 inanimate slaves toiled for civilization. Most of these mechanical subordinates worked in the industrial world. In fact, Fuller's map showed that the United States had "54 percent of the energy slaves, an army of 20,000,000,000."

In 1969, Fuller asked J. François de Chadenedes, a petroleum geologist, to figure out what it cost Mother Nature to make one gallon of petroleum. The innovator instructed the geologist to include the cost of photosynthesis as well as the slow cooking by heat and pressure into crude over millions of years. De Chadenedes obliged, estimating the price at more than $1 million per gallon. At the time, Fuller noted that the average American car owner consumed 300 gallons of petroleum a year. That meant each American burned through $300 million worth of natural capital with little to show for it. Fuller called it a cosmic crime: "Everyone knows that we should live on our [energy] income and not our savings account." But slaveholders are not known for their scruples.

Almost three decades later, Jeffrey Dukes, an ecologist, conducted a similar study. He calculated that every gallon of gasoline burned in a vehicle required the pumping or excavation of 98 tons of prehistoric buried plant material. "Can you imagine loading 40 acres of wheat—stalks, roots and all—into the tank of your car or SUV every 20 miles?" asked Dukes. He added up 1997 U.S. coal and petroleum consumption and concluded that it amounted to 97 million billion pounds of carbon, a number that translates into more than 400 times "all the plant matter that grows in the world in a year." Every day, in other words, the energy slaves of American car owners used the fossil-fuel equivalent of all the plant matter that grew on land and in the oceans that whole year.

The density of inanimate slaves and the extravagance of their owners reflect not only the cheapness of mechanical slaves' primary fuel but the legacy of human slavery. The values of one energy system have been neatly imposed on the other. There are only two ways to climb the ladder of comfort and material wealth in a civilized society: "One is to subjugate your fellow man and force him to produce an energy surplus that you appropriate," wrote Earl Cook, a Texas geologist, in a perceptive 1976 book. "The other is to harness the energy resources of nature in such a way that a surplus is produced." That fuel energizes more mechanical slaves to produce more surpluses. Cook doubted that industrial society could have happened in a society without slavery; slavery laid down a foundation "in which the primary energy flow was controlled so that a surplus was assured for the managers." Going from human slaves to inanimate ones was a slam dunk.

The deployment of energy slaves fastidiously follows what energy analyst Peter Tertzakian calls "the First Principle of Energy Consumption": as any Roman or Recife planter could have explained, the wealthiest people use the most energy. The average American now lives in a house bigger than a tennis court. The United States' wealthiest citizens live in McMansions, villas ten times that size. Their bathrooms are bigger than most shanties.

Human and petrol-based slave systems share other similarities. In the human slave world, scarcity bred respectful use; cheap abundance fostered contemptuous waste. Any energy system, human- or carbon-based, is really about commanding surpluses for convenience and easy living. Yet, over time, the concentrations of energy required to consistently deliver these creature comforts construct social pyramids that crumble when slaves—or petroleum—get too expensive. So far, we have rarely experienced scarcity of our cheap oil slaves, and we abuse them every day. But any dominant energy system thrives on its own inertia and creates a cognitive dissonance that causes good and often very smart people to rationalize shocking behavior. (Aesop, the blind slave turned storyteller, once noted, "Any excuse will serve a tyrant.") Just like slavery numbed Brazil, oil-based machines have locked decision making and thwarted innovation in the United States. And too much energy, as we have seen, can change the very metabolism of the master, if not his soul. Last but not least, every energy system creates its own startling dependencies and unpredictable dynamics. "If you put a chain around the neck of slave," Ralph Waldo Emerson once remarked, "the other end fastens itself around your own."

One of the remarkable characteristics of slaveholders as a class of people is their indebtedness. When soils have been mined by slaves for quick profits, a new reality follows. Surpluses decline, taxes increase, and the creditors come calling. Most Brazilian planters lived beyond their means, because it cost a lot to feed, clothe, and organize a household of obedient subordinates. The average North American, who must pay rising fuel bills to sustain an army of eighty-nine energy slaves, sings a similar lament today.

In his erudite account *Debt: The First 5,000 Years*, David Graeber, a British financial analyst and anthropologist, notes that people often sold *themselves* to pay their debts in ancient times. He doesn't think Americans behave much differently. "The great social evil in antiquity, the thing that Sharia law and medieval canon law were trying to ensure never happened again, was the scenario in which a family gets so deep in debt that they are forced to sell themselves, or sell their children, into slavery," says Graeber. North Americans now live on credit to support their own energy slaves and to buy largely unnecessary goods created by other energy slaves. "You have a population all of whom are in debt, and who are essentially renting themselves to employers to do jobs that they almost certainly wouldn't want to do otherwise, to be able to pay those debts. If Aristotle were magically transported to the U.S. he would conclude that most of the American population is enslaved, because for him the distinction between selling yourself and renting yourself is at best a legalism... We've managed to take a situation which most people in the ancient world would have recognized as a form of slavery and turned it into the definition of freedom."

In 1963, Alfred René Ubbelohde, a British professor of thermodynamics, warned that the uncontrolled proliferation of energy slaves would create big problems. In particular, he noted that the multiplication of energy slaves challenged "the political and administrative structure of a state" in three distinct phases. In the first phase, everyone hailed the machines as liberating instruments. But this pleasant chapter never lasted long. During the next phase, petroleum traders, machine owners, and labor unions argued over who should make more consumable goods: human beings or (efficient) machines. In the final stage, the sheer number of energy slaves would outstrip the ability of the state or local communities to manage them. "Is it possible that in some parts of the world we may already be nearing the social and genetic limits to making man the master, and inanimate energy the slave?" he asked.

Ubbelohde raised other pertinent issues too. If the costs of fossil fuels to feed these inanimate slaves steadily rose, then the whole machinery of modern life would stagnate. It was conceivable that a society might reach a saturation point for energy consumption and decide to regulate the density of energy slaves in its midst. The multiplication of energy slaves performing social tasks might also create political tensions, Ubbelohde reasoned—energy slaves did their work so unquestioningly that he compared them to "silent voters." They affected political and economic affairs as profoundly, he suspected, as had human slaves in ancient Rome. "Until proper recognition is given to true costs and true benefits resulting from an increased population of energy slaves," he wrote, "no programme to improve the material standard of life can be soundly based."

That recognition has yet to come. Tens of billions of inanimate slaves now rule our daily lives and contribute to our growing debt loads. Oil, the world's first trillion-dollar industry, feeds the majority of these busy servants, and BP and Shell see no problem that more energy slaves cannot solve. When oil was cheap, every nation was encouraged, as the American geologist H. Foster Bain wrote in *Foreign Affairs* in 1928, to "accommodate themselves to the presence of the new power" and welcome "the invisible slave who works but does not eat." But in so doing, we have allowed the master resource and its obedient machines to create a dangerous new servitude. Around the world, oil and its servants have reengineered the nature of government, the scale of science, the weight of populations, the metabolism of cities, and the purpose of economics. They did so in the methodical way slave-based plantations uprooted tropical forests but on a Walmart-like scale never before witnessed.

The subordination began in the cradle of slavery: agriculture.

5

The Unsettling of Agriculture

. . . . .

"For we know that among all nations alike
the master has the power of life and death over his slaves,
and whatever property is acquired by a slave
is acquired by his master."

GAIUS, *Institutes*, AD 161

FOSSIL FUELS UNSETTLED agriculture in much the same way that the automobile reconfigured American cities. Beginning in the United States and moving outwards, oil took a family enterprise that ran on sun, animals, and muscle and replaced it with machines operating on industrial-scale landscapes owned by a few corporations. It turned the long, often complex traditions of husbandry into an odd collection of chemistry, economics, and high energy spending.

Making food production dependent on hydrocarbon-fed inanimate slaves yielded, at first, two highly significant crops: it put more food on the table, and it poured an astounding number of rural people into cities. As farming, once a renewable enterprise, morphed into a giant mining operation governed by fertilizers, pesticides, and machines, it became impossible to sit down to a meal in many parts of the world without literally eating oil. The industrialization of agriculture created new scales of waste in the form of eroded soils, polluted watersheds, unemployed farmers, and piles of discarded food. Petroleum enabled factory farming to colonize many low-energy farming systems in Africa and Asia as well. And cheap oil ushered in a startling change in energy behavior: industrial farming poured more energy into agriculture than it got back in high-yield crops. Lastly, it disconnected people from the earthy basics of life. Many urbanites now regard food as a well-packaged product sold in standardized box stores that dispense calories the way a gas station sells fuel. "The industrial eater is, in fact, one who does not know that eating is an agricultural act," writes farmer and social critic Wendell Berry, "who no longer knows or imagines the connections between eating and the land, and who is therefore necessarily passive and uncritical—in short, a victim."

To appreciate oil's mastery of farming and the slave-like role of the industrial food consumer—like confined livestock, the urban eater swallows what industry places before him—requires a brief primer on farming. Agriculture was, after fire, the world's first energy innovation. When it took off nearly ten thousand years ago, it created a truly Promethean revolution. Until then, hunters and gatherers spent a quarter of their time collecting food surpluses from streams, oceans, or forests. Agriculture promised less volatility by organizing people and plants to harvest energy from the sun. The cultivation of rice, corn, and wheat, along with the domestication of poultry, cattle, and pigs, allowed early societies to create a surplus of calories in one dense spot. The bounty swelled human numbers like a field of pumpkins. The treasured surplus created by these plant-energy convertors changed society the same way petroleum would upend the United States. Because of their ability to capture sunlight and transform it into carbohydrates, wheat, corn, and rice served almost as the hydrocarbons of their day and drove the world's first major population booms.

But this burst of activity came with some extreme costs. Large cereal crop monocultures required not only irrigation but dedicated human workers—masses of slaves. And managing so much shackled energy required sizable armies and an authoritarian state. The rewards generally accrued to the 1 percent. As such, the agricultural revolution dramatically changed Mesopotamia, Egypt, Greece, and Rome. "Agriculture was not so much about food," explains U.S. writer Richard Manning, "as it was about the accumulation of wealth."

This new energy revolution also changed gender roles, especially in societies that adopted the plow for wheat, barley, rye, and wet rice crops. The tool required hefty muscle and bursts of speed that favored the male physique. Men in plow-based communities worked in the fields while women took care of the home. Groups that still used hoes and digging sticks, a vastly superior approach in pure energy terms, valued women's contribution in the fields, and their descendants exhibit greater gender equality today, with more women in politics, the workforce, and entrepreneurial activities. As the use of the plow shifted societies from matriarchal to patriarchal, noted the great French historian Fernand Braudel, "the reign of the all-powerful mother-goddesses" was supplanted by "the male gods and priests who were predominant in Sumer and Babylon."

Anthropologist Jared Diamond thinks agriculture might well have been, as he titled an essay, "the worst mistake in the history of the human race." Over time, it dramatically narrowed diets and created "diseases of civilization" such as diabetes, coronary heart disease, and obesity. Agriculture concentrated wealth—energy surpluses—in the hands of a few kings and pharaohs. It enslaved millions to convert solar-fed plants into human fuel. Early agriculture also spread infectious diseases acquired from domesticated animals throughout the cities that grew dense on irrigated food. The quest for more energy from impoverished wheat monocultures shortened life spans, weakened teeth, lowered bone density, and reduced people's stature. (Most hunter-gatherers were a good five to nine inches taller than early cereal-food growers.) "Forced to choose between limiting population or trying to increase food production, we choose the latter and ended up with starvation, warfare and tyranny," wrote Diamond in his 1987 essay.

But Diamond overstates his case. Take the example of China. Between the eighth and twelfth centuries, China constructed a unique agro-energy empire, diverting enough natural flows of energy to feed nearly half a billion people. It did so by carefully marshaling solar energy in intensely farmed plots of millet and wheat in the north and of rice in the south. Innovations in rice farming—the artificial flooding of land and multiple cropping—tripled the yield of an average peasant family. One square mile of carefully tended land could feed 225 peasants. Peasants hoed, fertilized, and irrigated these highly nutritious crops like some great garden. The land was manured with human shit. As French energy historians Jean-Claude Debeir, Jean-Paul Deléage, and Daniel Hémery note, "China avoided long-term energy shortages because the performance of its energy structures was unequalled anywhere; its dynamism was exceptional in history."

China's durable solar-farming system made its peasants self-sufficient and self-reliant. The system depended on masses of attentive gardeners. To maximize the area for plant growth, people even limited the use of grass-eating draft animals. As long as the food energy derived from crops could sustain its human workers, the system thrived. But as China's population grew from 100 million in the twelfth century to 500 million in the eighteenth, the country experienced a series of energy shocks. It ran out of both cultivable land and its primary source of heat: wood. Meanwhile, the human numbers needed to convert solar energy into food increased, while the surplus provided by the crops declined. Political uncertainty and instability prevailed, and China fell into a sort of Dark Age just as Europeans were mastering fossil fuels. Twentieth-century Chinese Communists tried to reenergize the old system with mechanical water pumps, scientific farming, and specialized seeds, but these reforms only eroded the soils, decimated forests, and aided the growth of deserts. The essential structures of the Chinese energy system remained intact until the 1970s, when industrial agriculture and urbanization assaulted the ancient model. Fossil fuels, chemical fertilizers, and dams then uprooted 400 million peasants.

The Industrial Revolution had the same effect on Europe's traditional agricultural energy system. By the time coal was discovered, famine, climate change, and disease had taught farmers to value diversity in both crops and animals and to prize small endeavors. Agricultural and peasant societies in Russia, Latin America, and India had also achieved remarkable ecological accommodations with the land, sharing both energy and wealth. In England, farmers and their laborers, along with squatters, managed arable lands, common grazing areas, and the wild commons they tellingly called "waste," because no solar energy was captured there. Monastic communities empowered villages with water mills and windmills. But their communal engagement and renewable solar-energy sources stood as obstacles to the expansion of cheap hydrocarbons. The purveyors of coal began their assault by introducing notions of private property. New land rules in England and across continental Europe effectively ended communal energy production and the very idea of a commons. Thrashing and seeding machines invaded the countryside. A series of revolts and riots convulsed the rural landscapes of eighteenth-century Europe. (On the same scale, protests now consume much of rural China, where millions have been uprooted and packed off to cities.) But the machines and their hydrocarbon feedstacks prevailed as resolutely as had slavery in Rome. Even Karl Marx, an urban historian with little regard for peasants was astounded by Europe's initial transformation: "Subjection of nature's forces to man, machinery, application of chemistry to industry and agriculture, steam-navigation, railways, electric telegraphs, clearing of whole continents for cultivation, canalisation of rivers, whole populations conjured out of the ground—what earlier century had even a presentiment that such productive forces slumbered in the lap of social labour?"

Perhaps the most significant chain shackled to farming by fossil fuels came from the Haber-Bosch process. In 1913, Fritz Haber, a patriotic German chemist, figured out how to draw nitrogen from the air by using a complex, energy-intense coal-based system to make ammonia. The metallurgist Carl Bosch, at BASF, later worked out a cheap commercial process. Before Haber and Bosch could produce any nitrogen for use on crops, the German government took over the factory for munitions production, thereby adding a few years onto World War I. But synthetic nitrogen fertilizers, as U.S. terrorists would later demonstrate, make powerful explosives. Haber also developed chlorine gas and the insecticide Zyklon B. The Nazis deployed the latter to kill millions in concentration camps. (Haber died in 1934, after the Nazis stripped him of his university position because he was a Jew.) In a reversal of the earlier development history, after World War II, U.S. munitions factories turned themselves into fertilizer operations in an attempt to expand nitrogen markets.

The liberal application of artificial fertilizers made from natural gas (it takes 33,500 cubic feet of methane to make one ton of anhydrous ammonia fertilizer) detonated human population numbers through higher crop yields. In 1900, most wheat and corn fields still depended on manure, guano, or saltpeter for a nitrogen kick. Although nitrogen makes up 80 percent of the atmosphere, it's not readily available for plants. When civilizations depleted their soils of nitrogen, they faced famine or mowed down more forests to make new cropland. But the Haber-Bosch process erased these boundaries. By 2000, almost every major crop, from wheat to soybeans, had tripled its yield, thanks to generous nitrogen subsidies. U.S. corn harvests rose fivefold; Japan's rice yields increased by three times. The combination of oil-based herbicides and scientifically engineered crops helped boost production even further. Today, the Haber-Bosch process feeds more than one-third of the world's population and accounts for half the nitrogen in every human body.

But this fast and furious plant growing changed the metabolism of both humans and the earth itself. Most vegetables and cereals now contain fewer proteins, minerals, and vitamins than they did one hundred years ago. Researchers suspect that the speed of plants' growth has diluted their uptake of good nutrients. Highly fertilized crops have also compromised the world's nitrogen cycle. Every year the planting of legumes, the spraying of fertilizers, and the release of nitrogen oxides from tractors and other combustion engines converts more nitrogen into reactive forms than is created by Mother Nature. Crops take up only 30 percent of the applied fertilizer, and the rest washes away. Scientists estimate that modern farming is leaking three times more nitrogen into the oceans, waterways, and atmosphere than they can absorb. This toxic leakage contaminates groundwater with nitrates and creates dead zones in oceans, lakes, and rivers. The doubling of fixed nitrogen has worsened the greenhouse effect, weakened the ozone layer, thickened smog layers, intensified acid rain, and poisoned vast expanses of water, from the South China Sea to the Gulf of Mexico, with blooms of nitrogen-loving creatures that gobble up oxygen supplies. Evan D.G. Fraser and Andrew Rimas, the authors of *Empires of Food*, note that the Haber-Bosch process "swapped out dependence on nitrogen for a dependency on the process to make nitrogen, which like so many elements of the modern world, is entirely reliant on fossil fuels." In 2009, some of the world's most prominent scientists warned in the journal *Ecology and Society* that "the growth of fertilizer use in modern agriculture" had transgressed the boundary for "human interference with the global nitrogen cycle."

AS FOSSIL FUELS put more people on the planet, their energy machines vacuumed farmers off the land. Raising corn by hand takes about three thousand hours of labor per acre; a machine can do the same work in twenty-seven. Between 1830 and 1930, the work of John Deere "singing plows," grain combines, threshing rigs, and gasoline tractors reduced the number of person-hours to produce one hundred bushels of wheat by 90 percent.

As the Americans exported their cheap-oil model for farming through Norman Borlaug's Green Revolution, the unsettling of rural people became a global epidemic. But the exodus really started in the United States. Nearly sixty years ago, historian David M. Potter observed in his book *People of*

Plenty that in 1820, nearly three-quarters of all Americans worked the soil. By 1950, that number had been reduced to 12 percent. Today it stands at a perilous 1 percent.

In 1906, civil rights crusader Mohandas Gandhi offered a powerful defense of the low-energy agrarian ideal. Gandhi, who opposed rapid industrialization and urbanization as fervently as he did British occupation, considered small, self-sufficient villages an enduring model for right livelihood. The village, he said, preserved a local economy, eschewed materialism, and served as a check on power. He noted that his forebears had known much about machinery but opted not to adopt some tools for fear that "we would become slaves and lose our moral fibre." India's ancestors, he said, "saw that our real health and happiness consisted in a proper use of our hands and feet." The mechanization of life in India would be a disaster for the nation, Gandhi argued. "It would be folly to assume that an Indian Rockefeller would be better than the American Rockefeller. Impoverished India can become free, but it will be hard for any India made rich through immorality to regain its freedom."

Rural resistance movements also appeared in the United States. In 1930, a group of Southern farmers and poets issued a manifesto challenging the mechanization of farming. In a collection of essays called *I'll Take My Stand*, the agrarians declared that standardized products, tractors, and fertilizers had "enslaved our human energies." Furthermore, the indiscriminate application of labor-saving machines had debased work, "one of the happy functions of life." Machines did not emancipate farmers, they said, but evicted them from the land. The agrarians condemned "economic super-organization" as a form of Communism and a threat to both artistic and religious ideals. They hailed the husbandry of the soil as the noblest of vocations.

One of the first British critics to catalogue the impact of fossil fuels on agriculture was Sir Albert Howard, a soil researcher who studied farming in India for thirty years. In *An Agricultural Testament*, still one of the best books ever written about farming, Howard decried the expansion of farm size and the shrinkage of farm populations. He considered the replacement of animal and plant compost with artificial fertilizers to have been a disaster and the overtaking of small mixed farms with machine-managed monocultures to be ominous. "Engines and motors of various kinds are the rule everywhere," he wrote. "The slaves of the Roman Empire have been replaced by mechanical slaves. The replacement of the horse and the ox by the internal combustion engine and the electric motor is, however, attended by one great disadvantage. These machines do not void urine and dung and so contribute nothing to the maintenance of soil fertility. In this sense the slaves of Western agriculture are less efficient than those of ancient Rome." The flood of cheap food caused by the infusion of fossil fuels had already come at great cost in "the steady growth of disease in crops, animals, and mankind." A people fed on improperly grown food, he predicted, would ultimately need "an expensive system of patent medicines, panel doctors, dispensaries, hospitals, and convalescent homes."

But Howard saw only the beginning of oil's corrosive impact on agriculture. After World War II, farms got bigger, monocultures expanded, and countries started to build empires of livestock. In 1970, *National Geographic* detailed this transformation in glowing terms. Although more than 650,000 farmers abandoned rural America every year, the magazine said, farming could only get better and more efficient. Machines picked tomatoes bred for machine picking. One man in an air-conditioned tractor could harvest a crop of corn that once would have required eighty men. Americans could now eat strawberries in January; scientists had planted heating cables in the ground to warm the soil for the novelty of asparagus in December. More than 40 percent of the stuff sold in grocery stores hadn't been available a dozen years earlier, the article gloated. A factory on wheels could mow down rows of celery. "Because only one person in 43 is needed to produce food, others can become doctors, teachers, shoemakers, and janitors," said the magazine. And the one man left on the farm could now raise 100,000 broiler chickens all by himself.

As more oil poured into U.S. agriculture, the nation's vocabulary changed. Speeches by agricultural leaders in the 1970s talked about "agripower" and the total control of "inputs and outputs." The goal was no longer to feed Americans but to "generate agridollars through agricultural exports." The average size of farms had grown from 195 acres in the 1940s to more than 400 acres. Bigger was better, and farmers who couldn't get big enough would have to get out.

Around the same time, U.S. sociologist Fred Cottrell noticed a funny thing about this expanding petroleum-based food production system: it consumed more energy than it produced. Hoed corn cultures created greater value than did the United States' mechanized corn growers, for example, and muscle power in the Philippines produced higher rice yields than did machines in Louisiana. Although U.S. corn yields had risen from forty to ninety bushels an acre between 1950 and 1970, Cottrell wrote, the cost of energy inputs had grown even faster, in the form of fertilizers, "acre-eating" machinery, and herbicides.

Cottrell attributed these rising energy inputs to several factors. For starters, it took lots of oil to reduce the amount of time people spent in the fields. Heavy tractors, seeders, and combiners required more energy to make, maintain, and repair. And bigger farms required more mechanical work. Cottrell concluded that a high-energy system based on cheap oil didn't behave normally at all: "What the industrialist demands from the farm is not the maximum energy he can secure, or necessarily any labor force; it is food itself, in sufficient amounts to maintain the industrial population and assure its growth." He wasn't alone in being dismayed by this waste. According to energy analyst Earl Cook, in 1973, to deliver 3,300 kilocalories of food, it took 26,745 calories of fossil fuels in plowing, processing, transporting, and storage: more than eight times the energy available in the food itself.

Ecologists Eugene and Howard T. Odum, also writing in the 1970s, called U.S. oil-based farming ruinous due to its ungainly scale and its enslavement to unsustainable energy flows from petroleum. "The fields were sown by machinery, tilled by tractors, and weeded and poisoned by chemicals. Epidemic diseases were kept in check by great teams of scientists in distant experiment stations," wrote Howard Odum. But the high yields were an illusion, he said, because the average person had no idea how much fossil fuels subsidized the affair: "A whole generation of citizens thought that higher efficiencies in using the energy of the sun had arrived. This was a sad hoax... People are really eating potatoes made partly of oil."

IN THE DECADES since then, as a 2010 energy audit released by the United States Department of Agriculture confirmed, U.S. potatoes have become increasingly oily. In the 1960s, the food system ate up 12 percent of the nation's energy budget; in 2007, it gobbled up 16 percent. The USDA audit explained how with a brief portrait of lettuce production in California. A precision seed planter attached to a gasoline-powered farm tractor plants the lettuce. Then a diesel-powered broadcast spreader applies nitrogen-based fertilizers, pesticides, and herbicides. Electric-powered irrigation equipment transports water to the plants. The farmer drives to the store for supplies. Field workers imported from Mexico harvest the vegetables in boxes, which are then loaded onto a gasoline-powered truck, which drives to a processing plant. More machines sort, cut, and clean the lettuce. The packaged lettuce is loaded into a refrigerated truck or railcar destined for an East Coast grocery store, where it is promptly placed in a refrigerator powered by coal-fire-generated electricity. A motorist purchases the lettuce, drives it home, and places it in the fridge. In the end, half of the lettuce is thrown out and hauled by truck to a landfill. Raw food now travels, like winter tourists, an average of 1,500 miles, processed food about 1,346 miles. The USDA report, as dry as seed, concluded that "dependence on energy through the food chain raises concerns about the impact of high or volatile energy prices on the price of food, as well as domestic food security and the Nation's reliance on imported energy."

In 2005, energy consultant Jean-Marc Jancovici examined the modern state of farming in France. As in the United States, Jancovici found, industrial farming accounted for roughly 20 percent of energy spending. Jancovici converted those kilowatts into calories, or "slave equivalents," to calculate the energy used by transportation, industry, the heating of buildings, and farming. He estimated that the mechanical power of tractors and harvesters represented twenty slaves per French citizen. Another twenty-six per capita toiled for industrial services. Overall, Jancovici estimated, modern energy consumption gave each French citizen one hundred invisible personal slaves.

This kind of convenience, however, comes with a ball and chain. One striking aspect of fossil-fueled farming has been its relentless concentration. Petroleum has done to farming what Standard Oil did to the oil refiners: put more power into fewer hands. Three firms now control the entire U.S. meat-packing business. Just 2,100 feedlots feed 90 percent of all cattle. Four firms manufacture 60 percent of the nation's chickens in factories. One company, Monsanto, controls most of the world's sales of corn and soybean seed. Four firms, including Kraft and Saputo, control most of the dairy processing in the United States. Wal-Mart, the ExxonMobil of cheap things, has monopolized 30 to 50 percent of all U.S. retail food sales. Campbell's occupies most of the grocery-store shelf space for soup. Frito-Lay sells half of the sodium-rich corn and potato chips.

Standardization, another petroleum trait, has turned farming into highly specialized and fragile monocultures. Corn used to come in 50 varieties suited for different locales and resistant to particular pests, but agribusiness prefers just 6. Wheat, which once boasted 30,000 varieties, now comes in only three or four specialized types that must be engineered to withstand 300 pounds of pesticide poison per acre. The once mighty community of 5,000 potato varieties has largely been reduced to the Burbank, a tuber suited to making french fries, which fatten both North American diners and fast-food profits. Most of the apple's 7,000 varieties are gone. Lettuce, once seen in a global array of sizes, shapes, and tastes, is now represented mainly by two types: iceberg and romaine. The UN's Food and Agriculture Organization estimates that three-quarters of the genetic diversity in the world's food crops has been washed away, dug up, or paved over, despite earnest attempts to preserve diversity. Rusts, blights, and insects have had a field day with one-crop farms. Cornell University ecologist David Pimentel estimates that half of the world's annual crop yield gets eaten by 70,000 different pests and plant diseases, despite an annual expenditure of $35 billion on pesticides.

Livestock diversity has suffered similar erosion due to petroleum's preference for homogenization. Eighty percent of all dairy cows are Holsteins, and 60 percent of all beef cattle are Angus. Forty percent of all sheep are Suffolk. Pig factories prefer the English Large White. In the last 100 years, one in 6 livestock species, each carefully bred to withstand local calamities, has gone extinct. Sixty breeds of cattle, goats, horses, and pigs have vanished since 2007 alone. About 1,350 of the world's remaining 6,500 livestock breeds now face the same fate as their rural caretakers. Animals bred to resist disease, endure drought, and survive cold are being replaced with dumbed-down, high-yield factory "units" shaped by petroleum inputs consisting of vaccines, antibiotics, and formulated and highly fertilized feeds. About 70 percent of the world's chickens and nearly half the world's pigs and cattle are now raised in confined factories totally separate from the land. High-yielding grains helped create this so-called livestock revolution. Since 1961, the number of chickens, ducks, and turkeys on the planet has quadrupled from 410 million to 15.7 billion. Cattle and pig populations now exceed a billion each. This industrial system accounts for 8 percent of the world's surface water use, 18 percent of the world's greenhouse gases, 20 percent of the global animal biomass, and 50 percent of all antibiotic consumption. It takes almost three-quarters of a gallon of oil to make a pound of beef.

Industrial farming has also, sadly, democratized the waste of food. About 27 percent of edible fruits, vegetables, oils, and dairy products in North America spoil in transport, rot in the fridge, age in a grocery store, or get thrown out at home. In England, food waste may be as high as 50 percent. A University of Arizona study found that the average U.S. family squanders about $2,275 worth of food a year. The amount of energy lost through rotting or uneaten food accounts for 2 percent of annual oil and electricity spending in the United States.

G.K. Chesterton predicted this outcome nearly a hundred years ago. Like Gandhi, Chesterton supported communities of small farmers because they produced wholesome food, made people happy, and did not ignore the importance of religious faith in everyday life. Nor did small farms concentrate economic or political power. Chesterton argued that the growing presence of inanimate slaves in the countryside would not only enslave people's thinking but ultimately mechanize their leisure time. "If our ideal is to produce things as rapidly and easily as possible," he wrote, "we must have a definite number of things that we desire to produce. If we desire to produce them as freely and variously as possible, we must not at the same time try to produce them as quickly as possible. I think it most probable that the result of saving labour by machinery would be then what it is now, only more so: the limitation of the type of thing produced; standardization."

Rural depopulation caused by the petrolization of farming in the United States has created a number of largely unrecognized cultural emergencies. For several decades Wendell Berry's elegant writings have highlighted these crises. The first has been the systematic abuse of rural areas by "itinerant professional vandals" in industry and government. "We now have more people using the land (that is, living from it) and fewer thinking about it than ever before," says Berry. The farming diaspora also disconnected people from place and home. "When people do not live where they work, they do not feel the effects of what they do." All high-energy-use countries, from Communist China to the United States, treat rural areas as dumping grounds for exploitive industries and nuclear waste, using the excuse that no one lives there.

Berry believes that the globalization of food production could well become the twentieth century's most ruinous legacy. The factory-farm model, a U.S. invention, has nearly wiped out local traditions, local farmers, and local crop diversity around the globe with the bold lie that factory techniques can conquer the quirks of Mother Nature. Writes Berry, "As soon as pests, parasites, diseases, climate fluctuations and extremes are left out, resistance to these things is also left out; and this resistance, in the soil and lives that come from the soil, is what we call health. And so for total control we have given up health."

The factory food system designed to fatten plants and animals with oil subsidies has also supersized people. Forty years ago the average U.S. citizen was fairly lean. By 1994, researchers had begun to notice a startling trend: Americans were growing chunky. Two decades later they called it an epidemic. Thanks to fast food, a diet rich in carbohydrates, and labor-saving energy slaves in the form of household appliances and automobiles, two-thirds of Americans are now overweight. Thirty percent qualify as obese. Nearly a quarter of all American children are corpulent. Visitors to the United States are often dumbfounded by the abundance of gargantuan people. Farmers in low-energy societies expended about 1,000 calories a day in the fields, consuming about 3,000 calories a day if they were lucky. That represents a 3 to 1 subsistence efficiency. Our inanimate slaves allow many of us to expend but 300 calories a day in physical work while we consume 2,100 calories: a subsistence efficiency of 7 to 1. Mechanical devices have so reduced physical activity since 1945 that the average daily energy expenditure of a person in England today has declined by 800 calories a day. In energy terms, that's equivalent to a 10-mile walk.

Factory farming has done something even more unprecedented in the West: it has fattened the poor most of all. Obesity, of course, has a long pre-petroleum history. With the advent of organized grain farming, it quickly became associated with wealth: a fat person symbolized abundant calories. One of the first depictions of obesity can be found in an Egyptian tomb that shows a lean slave feeding his fat owner. Today's processed foods, which contain refined grains, fat, and sugars, are not only cheap but deliver high energy fixes. Industry spends nearly $11 billion a year promoting fast foods and supersized meals. Fast-food diets are much more affordable than lean meats, fish, vegetables, and fruit. A 2004 study by Adam Drewnowski and S.E. Specter identified a rueful paradox: "Americans are gaining more and more weight while consuming more added sugars and fats and are spending a lower proportion of their income on food." Cheap oil has powered it all.

Public health authorities and other experts say low-fat diets and exercise are the cure. But these rote recommendations miss the true energy dynamics at play. The Centers for Disease Control and Prevention calls the problem of obesity "complex," and the fat epidemic has spawned a professional industry of fat watchers, diet gurus, and the like. But really, it's all about the energy. Whenever a society changes the metabolism of farming with fossil fuels, it clogs its own arteries. Petroleum-based food supplies deliver more carbohydrates to the table than did traditional agriculture. An average American now consumes about four bushels of wheat a year. These modern grains are much higher in carbohydrates than previous grain varieties, and too many carbohydrates affect insulin production in the body the way fertilizers impact the nitrogen cycle. Naturally skinny people can absorb some of these carbs, and the fat they store produces energy for later drawdowns. Fat people, however, can't draw down on stored fat, because carbohydrate consumption has made their insulin levels go haywire. "So fat people are fat not because they overeat—they overeat because they're fat," explains nutrition expert Michael Eades. Wherever aboriginal populations have returned to traditional diets low in carbohydrates, people have not only lost weight but have shed such health problems as diabetes.

Every culture in the world that has adopted U.S.-style farming practices and processed foods has also grown big and unhealthy. Some critics call the phenomenon "peak health." In particular, oil-producing countries have some of the most supersized citizens on earth, because they eat junk food and drive everywhere. In Mexico, Bahrain, Kuwait, and Saudi Arabia, 35 percent of the adults are considered obese, according to the Harvard School of Public Health. Kuwait, for example, has the highest rate of obesity in the Arab world; more than half of its women have large waistlines. A 2003 study found that 75 percent of the workers at the Kuwait Oil Company were overweight, with field workers weighing even more than office workers. The researchers, of course, recommended "an active lifestyle, healthy eating habits and weight control programmes." Qatar breaks all records. An astonishing 40 percent of its people are fat, and one in five has diabetes. Five years from now, 73 percent of Qatari women and 69 percent of the men are expected to qualify as obese. Doctors there perform fat-removing surgery on children. Most of Qatar's citizens have the same lousy health profiles as residents of the South Pacific, where American processed foods displaced diets of fish and vegetables.

IN THE 1920S, to slow the erosion of the United States' small farmers, who bought Ford vehicles in droves, Henry Ford championed ethanol as a better fuel source than oil. The magnate, who had originally planned to run the Model T on alcohol, thought biofuels could alleviate the economic stress in rural America by providing an additional cash crop. In 1925, Ford told the *New York Times* that "the fuel of the future is going to come from fruit like that sumac out by the road, or from apples, weeds, sawdust—almost anything." Ford also promoted a movement called farm chemurgy, which aimed to turn ethanol, soybean, and hemp into fuels and consumer products. Cheap oil nullified the initiative at the time, but when oil prices reached nearly $150 a barrel in 2008, the U.S. industrial food system turned again to the idea of ethanol made from corn. Across the ocean, Europeans proposed making biodiesel from palm oil raised on the Indonesian plantations that had replaced tropical rain forests. Soybean, switchgrass, algae, and sugarcane are other key biofuels. But the sudden rush by the United States and Europe to cannibalize crops for fuel dramatically raised corn, meat, and egg prices around the world. It also sparked food riots in places like Mexico and North Africa. One American agricultural technocrat praised the initiative but noted that it could force a trade-off between food security and energy security. Other energy experts calculated that completely replacing fossil fuels with biofuels would require the harvest of all farm, forest, grassland, and aquatic plants on the planet. Researchers Mario Giampietro and Kozo Mayumi concluded that "full substitution of fossil energy with agro-biofuels is impossible."

It is an old debate. In the nineteenth century, Josef Popper-Lynkeus, an Austrian physicist and engineer much admired by Albert Einstein, analyzed the cost of replacing coal for domestic heating with alcohol made from potatoes. The geography needed proved to be ungodly. When proponents argued that the potatoes could be grown in greenhouses instead,

Popper-Lynkeus pointed out that greenhouse farming in Europe consumed so much coal that any energy returns from the resulting potato alcohol would be paltry.

David Pimentel has raised the same questions about ethanol. Given the fuel's meager energy returns—the scientist says they are negative; proponents insist that ethanol provides twice the amount of energy invested in its production—Pimentel questions the use of biofuels on even a limited scale. Biofuel advocates, he says, ignore the fact that 60 percent of the world's population is malnourished and too much land has already been converted to industrial uses. "When these people talk about biofuels providing us with our energy, they need to look at the facts right now," said Pimentel in a 2006 interview. "Eighteen percent of all corn is going into ethanol production. We're getting 4.5 million gallons of ethanol. That's 1 percent of U.S. petroleum use. It's 1 percent. If we use 100 percent of U.S. corn, and we won't do that, but if we used 100 percent, what would that do for us? Six percent. And ethanol is being subsidized at 45 times the rate of gasoline."

The industrial food system, which enslaves nearly 90 percent of all cultivable global land to the production of grains for export, has provoked a growing rebellion around the world. As oil prices increase, many peasants and small farmers are abandoning the so-called Green Revolution for old-fashioned husbandry. In the United States, the sustainable farm movement has challenged Big Ag by cultivating smaller and more diverse plots with natural manures and fewer machines. A growing number of citizens are eating more "slowly" and making a point of buying food raised locally. "We do not need to end up in a food dictatorship and food slavery," says Vandana Shiva in *Soil Not Oil*. "But to remain authentic, organic farming must be biodiverse, it must stay in the hands of small farmers, and it must deepen food sovereignty."

Wherever U.S.-style petroleum plantations have conquered traditional farming, people now experience diminishing returns or rising energy costs. Although Norman Borlaug promised India's farmers more food, his recipe of high-yield grains and chemical fertilizers has impoverished the farmers as well as the soil. Shiva, a former nuclear physicist turned agrarian reformer, has documented the loss of micronutrients such as iron, copper, and magnesium in the Punjab. Half of the region's soil samples also show zinc deficiencies. "The productivity of wheat and rice has been fluctuating and even declining in most districts in Punjab, in spite of increasing levels of fertilizer application," she writes.

Bangladesh, where half of the people still work the fields, also spends more energy for less food. Between 1990 and 2005, the country increased cereal yields from 34 to 46 million tons, a jump of 35 percent. But it took a 50 percent increase in energy in the form of machines, fertilizers, oil-powered irrigation, and pesticides for Bangladesh to achieve these gains. In 2010, a group of researchers led by Shaikh Khosruzzaman looked at the growing energy intensity of the country's farms, which increased tenfold between 2000 and 2008, and suggested that employing more petroleum slaves at the expense of humans and animals had been counterproductive. "The increasing trend of energy intensity in the agriculture sector of Bangladesh does not support sustainable development," they concluded.

No one knows for certain how industrial farming will fare as its petroleum subsidies grow ever more expensive or what will happen to factory farming as climate change dries out some regional monocultures while flooding others. But Cuba has given the world a glimpse of the possibilities.

In exchange for sugar, orange juice, and nickel, the Soviets had provided the island with 90 percent of its oil and nearly 60 percent of its food. But in 1991, the Soviet state fell apart, and shortly afterwards the United States embargoed the island. The two events left Cuba with no fuel for its tractors and no fertilizer or pesticides for sugar and tobacco crops. The island experienced an oil price shock that economists compared to a crashing plane. Cubans called it the "Special Period." At the time, small farmers worked only 12 percent of the land base. A nation that had gotten the majority of its food from foreign countries went on an extreme diet. The calorie intake of the average Cuban dropped from a healthy 2,600 a day to a lean 1,000. Many Cubans shed as many as 30 pounds. Thousands went blind from malnutrition while pregnant women suffered anemia.

In response, the Cuban state turned to small farmers and entrepreneurial *organopónicos*, or small urban plots, to fill the food gap. Today, 81,000 acres of these urban plots employ thousands of people. (A typical 1.7-acre lot might employ 25 people full-time.) "What happened in Cuba was remarkable," says Laura Enriquez, a U.S. sociologist. "The Cubans went for food security and part of that was prioritizing small farms." With no petroleum-based fertilizers or pesticides at hand, the Cubans created 170 compost centers where wriggling worms now make 9,300 tons of soil a year. The country also set up centers to make natural pesticides such as verticillium and *Beauveria bassania*. Oxen replaced tractors, and people raised rabbits and chickens on rooftops. In short order, these small enterprises ended the food crisis, providing a rich stream of organic spinach, onions, chives, garlic, and tomatoes. Small farmers now grow most of Havana's fruits and vegetables. The Cuban government, which replaced its "Homeland or Death" slogan with "A Better World Is Possible" during the crisis, also allowed small farmers to keep 50 percent of their profits. It learned that campesinos were far more resilient than state-run agricultural enterprises or the socialist equivalent of agribusiness in North America. The effective use of land ably proved "the capacity of small farmers' methods of production and organization to contribute to the national food balance, even with scarce external inputs," reported Cuban researcher Fernando Funes-Monzote. The Cuban diet, once dominated by rice, beans, and pork, has become more diverse and vegetable-rich. In the end, what Peter Rosset and Medea Benjamin called the "the largest attempt at conversion from conventional agriculture to organic or semi-organic farming in human history" upheld the veracity of Thomas Jefferson's abandoned American ideal: that "small land holders are the most precious part of the state."

The last word goes to G.K. Chesterton, who championed small-scale food production back in the 1920s. "The quickest and cheapest thing for a man who has pulled a fruit from a tree," wrote Chesterton, "is to put it in his mouth. He is the supreme economist who wastes no money on railway journeys. He is the absolute type of efficiency who is far too efficient to go in for organization. And though he is, of course, an extreme and ideal case of simplification, the case for simplification does stand as solid as an apple tree."

6

The Viagra of the Species

. . . . .

"Can you think of any problem, on any scale,
from microscopic to global, whose long-term solution
is in any demonstrable way, aided, assisted
or advanced by having larger populations at the local level,
the state level, the national level or globally?"

ALBERT A. BARTLETT, "The World's Worst Population Problem," 1997

IN 1962, Marion King Hubbert, a Shell Oil geologist and one of the United States' greatest scientists, wrote a short treatise on energy for the U.S. government. The man who coined the term "peak oil" warned that humans had recklessly tapped into global energy flows. As a species, we had dammed rivers, harnessed the wind, tamed the atom, and exploited the stored energy in coal, oil, and natural gas created by the sun and plants 500 million years ago. Gorging on this remarkably rich energy buffet, humans had created a world of continuous growth. But the multiplication of humans and their inanimate slaves had upset "the ecological equilibrium in the direction of an increase in the human population."

In his report, Hubbert called this population explosion a wild departure from the norm. For nearly a million years, human numbers had increased so slowly that they doubled only once every 100,000 years or so. After the birth of Christ, the doubling time shortened to about 560 years. Nevertheless, wars, climate events, and plagues kept the human population at 300 million or less. Beginning in the sixteenth century, though, population graphs started to rise like ExxonMobil's profit margins. By 1820, the species had crossed a dramatic threshold and hit the 1 billion mark. Assisted by coal's mechanical slaves, people were living longer and reproducing more successfully. Global population took off at a rate of 2 percent a year and topped 3 percent in the United States. By 1900, the world's population had jumped to nearly 1.5 billion, and it doubled again by 1960. "These recent events have had no precedents in human history," explained Hubbert in 1962. He attributed the boom to the "progressive manipulation . . . of the large stores of energy contained in the fossil fuels."

Hubbert set out three possible futures for a species hooked on finite stocks of hydrocarbons. In the first scenario, with some technological prowess, the human population could stabilize on nuclear power. In the second, unbridled energy consumption could lead to a population overshoot and a dramatic collapse, followed by a lower standard of living. In the third, said Hubbert, "We could fall into a state of confusion and chaos, including nuclear warfare" and "suffer a cultural decline." The transition to a stable world was possible if politicians planned for it, Hubbert thought. But he would later reflect that the greatest obstacle to change was cultural: "During the last two centuries we have known nothing but exponential growth and in parallel we have evolved what amounts to an exponential-growth culture, a culture so heavily dependent upon the continuance of exponential growth for its stability that it is incapable of reckoning with problems of non-growth."

Until hydrocarbons entered the drama of demographics, human life, as novelist F. Scott Fitzgerald might have put it, was largely about "the pursued, the pursuing, the busy, and the tired." For most of human history, baboons outnumbered human beings. As Hubbert noted, the introduction of slave-based monocultures nearly eight thousand years ago created a brief population surge, and then the numbers stabilized again. When blessed with good weather, fertile soils, and sound government, humans were able to secure large amounts of surplus food and thrive. But when cursed with cold spells, famine, or inept rulers, early societies tightened their belts. To keep things on an even keel, many communities controlled their numbers by spacing births, killing infants, or enrolling sons and daughters into celibate orders or warrior castes.

Prior to 1750, humans generally, with some notable exceptions, lived off low amounts of energy, using whatever nature provided. Both agricultural and hunting societies tended to be highly fertile, and they placed great value on children (a source of energy and care) and on the extended family. Due to high mortality, the elderly were few but were respected for their wisdom. People ate more plants than animals. The family, a small welfare enterprise, not only ordered economic life but provided most goods and services. Low-energy cultures generally imagined the world as a cycle, never as an ascending growth line. They eschewed moneylending and tended to be fatalistic. They also prized a sense of place. Had the world not commercialized fossil fuels, demographer Graham Zabel estimates, the human population would not have grown much beyond 1 billion.

But an infusion of coal, followed by its steam-driven mechanical slaves, demolished the old demographic order.

European nations recorded the first dramatic population surges between 1500 and 1820. France and Italy experienced growth of between 50 and 80 percent; England's population jumped by 280 percent. Between 1800 and 1900, Europe's population grew like the amazing kudzu plant, from 187 million to 400 million people. Why this growth spurt? According to historian David Hackett Fischer, "an improvement in material conditions was part of the answer... Husbands and wives decided to have more children because the world appeared to have become a better place in which to raise a family."

If coal was the Spanish fly of the nineteenth century, then oil surely served as the Viagra of the twentieth. Historian J.R. McNeill notes that after 1950 (the year oil accounted for 20 percent of world energy consumption), the rate of global population growth increased at "roughly 10,000 times the pace that prevailed before the first invention of agriculture, and 50 to 100 times the pace that followed." McNeill offers some startling statistics. About 80 billion hominids have lived and died on earth over the last 4 million years, he writes. "All together those 80 billion have lived about 2.16 trillion years. Now for the astonishing part: 28 percent of these years were lived after 1750; 20 percent after 1900; and 13 percent after 1950. Although the twentieth century accounts for only 0.00025 of human history (100 out of 4 million years), it has hosted about a fifth of all human-years." According to current estimates by the United Nations Population Fund, the world's population reached 7 billion on October 31, 2011.

This demographic revolution, aided by New World crops, created a new set of energy values. The people on fossil fuels, perhaps the most narcissistic and bankrupt cohort in the history of the species, shop incessantly and genuflect to the market or the state, the provider of goods and services. They have fewer children. They prize small families or no families and primarily live in cities. They eat lots of animal protein, tend to be obese, and live as long as J.D. Rockefeller. They use condoms and support abortion. They accept women in the workforce, eschew marriage, hail secularism, promote individualism, and line up to be served by inanimate reproductive technologies. Highly mobile, they prize no place in general. Their elderly are numerous but neither valued nor respected. The overwhelming presence of mechanical slaves in everyday life has created the temporary illusion that children are not needed to care for the old.

EXPERTS HAVE long debated energy's role in this great human population convulsion. Many academics argue that the so-called demographic transition happened because of economics, vaccines, sewers, cotton clothes, and a profusion of comforts. They often cite China's incredible multiplication of peasants nearly five hundred years ago, which happened without fossil fuels. But others argue that that blip actually reinforces the primacy of energy: employing disciplined farming that closely resembled gardening, the Chinese secured more calories for their population, which in turn supported a hefty population increase. That's pretty much what has happened, on a much greater scale, with coal and oil. These finite stocks of energy have been manipulated using inanimate slaves to secure more calories for humans.

In particular, fossil fuels have driven the demographic revolution by boosting agricultural production. As noted, fertilizers, high-yield grains, factory-produced livestock, and billions of inanimate agricultural servants have put more food on the table. Although many people believe that food production must be increased to feed the hungry masses worldwide, human population grows or falls as a function of food availability, says Cornell researcher David Pimentel. Lemmings, rabbits, and rats explode in number whenever more food is available, and human beings are no different. Fossil fuels have not only increased yields but have made it easier to move crops around. The tractor, the chain saw, and the irrigation pump have also made it easier to industrialize vast landscapes for crop production. "By increasing food production for humans, at the expense of other species, the biologically determined effect has been, and continues to be, an increase in the human population," say Pimentel and Duke University researcher Russell Hopfenberg.

Along with this remarkable demographic boom came new ideologies. One of the most insidious was eugenics, the idea of building a master race. By the 1900s, many scientists, religious leaders, economists, politicians, and feminists in industrial nations from Germany to Canada were responding to the population boom by demanding the culling of the "unfit." Experts and professionals enthusiastically endorsed programs to sterilize, confine, or weed out the poor, the disabled, and the "feeble-minded" so that the so-called fit might prosper.

Catholic social critic G.K. Chesterton recognized the movement for what it was: an attempt to breed people like slaves for a mechanized life. "The shortest general definition of Eugenics on its practical side is that it... propose[s] to control some families at least as if they were families of pagan slaves," Chesterton wrote. Societies fueled by human slaves had no more use for the traditional extended family than did societies based on the energy slaves of fossil fuels. Some slaveholders tried to breed their slaves like cattle; others tyrannically controlled women's lives, including their lactation periods, and the treatment of sick children. The eugenics agenda remains an undercurrent of reproductive technologies today.

Fossil fuels have not only put more people on the planet; they have extended life expectancy and the "biological warranty." In 1930, the average man in the United States lived to age fifty-eight and the average woman to sixty-two; today, those figures are seventy-four and eighty. Centenarians are the fastest-growing segment of the U.S. population. In the process, oil and its inanimate slaves have turned a stage of life into a disease or institutional condition. Forecasts suggest that the average life expectancy in the United States could rise to eighty-eight by the end of this century, though declining energy spending and rising obesity rates would seem to challenge such optimism.

Yet petroleum's gift of long life has come with a price. In almost every fossil-fuel nation, the neglect and abuse of the aged in industrial-style institutions has generated uncomfortable debate. The many new age-extending technologies not only raise ethical issues but require more inanimate slaves and higher energy consumption. In *The Technological Bluff*, the influential French philosopher Jacques Ellul catalogued the physiological trade-offs of the so-called demographic transition for those in the developed world. He understood that living longer did not mean living better. "We have less resistance to grief," he wrote, "to fatigue, and to privation. We have less resistance to lack of nourishment, variations in climate, and internal and external stresses. We are more susceptible to infections... Our senses are less sharp, especially sight and hearing. Our nerves are much more fragile (we suffer more from insomnia and distress). We have to take more precautions and are more easily laid up by little things. We have more opportunities in life and live longer, but we live diminished lives and do not have the same vital force. We have to compensate for new deficiencies by artificial procedures that in turn produce other new deficiencies."

Demographers initially celebrated the population boom as a form of Promethean progress. They forecast that everyone around the globe would soon live in a nuclear family, with an average of one or two healthy children, and enjoy godlike life expectancy. But abrupt surpluses of energy in any population have rarely created stability, and the demographic ideal of low fertility and long life has rapidly become something of a social conundrum. Has human population now peaked? ask the people counters. "In the next fifty years," warns demographer Graham Zabel, "as the world's remaining oil resources are consumed, ... the world population could suffer a precipitous decline."

Over the past century, our reliance on fossil fuels has significantly changed fertility trends. Prior to the appearance of inanimate slaves, the average U.S. woman bore an average of eight children. Today that figure is less than two. Almost every society, from China to Iran, that has adopted petroleum-based living standards has opted for smaller families.

The proliferation of labor-saving machines meant that the family didn't need as many children to provide essential goods and services. As sociologist and energy specialist Fred Cottrell concluded nearly fifty years ago, "the decline in the Western birth rate was related to the way in which the products of increased energy were distributed." It was "sharpest among the households receiving the largest share of the increased output, and least sharp among households receiving the least share," Cottrell wrote. "Children in the urban home perform few functions which a machine cannot perform more cheaply; the mounting cost of urban housing, clothing, recreating, and education makes the cost of child-rearing very much greater than any financial gain which is likely to result from his services to the family of origin. Parents, therefore, increasingly seek means to escape the costs which child-rearing entails."

The demographic changes driven by fossil fuels in turn expanded the role of the state. Realizing there was no longer any guarantee "that care and affection lavished upon children [would] be reciprocated in the declining years of parents," noted Cottrell, citizens of Western societies turned to government institutions for help in the form of old-age security. The state forced the aged into retirement and warehoused them.

Dropping fertility rates have spawned a "birth dearth" in industrial nations today. To replace an existing population, each woman must have an average of 2.1 children. That's not happening in many European nations now, or in Japan or parts of North America. In Russia, fertility rates have reached all-time lows while mortality rates have climbed. Nations like Iran and Thailand also record low birth rates. The number of elderly people currently outstrips the number of children in Japan, Italy, Bulgaria, and Hungary. Some demographers fear rising energy prices will trigger a worldwide economic meltdown that could leave tens of millions of aging seniors languishing in industrial nursing homes while their children suffer from "long years of overtaxation, rising crime, and political instability." What will happen "if the economy and the welfare state shrink significantly?" asks U.S. anthropologist Stanley Kurtz. "Quite possibly, people will once again begin to look to family for security in old age—and childbearing might commensurately appear more personally necessary."

Population decline, something the designers of social security and socialized medicine never imagined, has several energy-related drivers. The movement of women into the workforce is one. In societies revving their engines on more oil, rising female literacy and falling female fertility go hand in hand. Urbanization has played an incendiary role. The fertility drop may also reflect widespread pollution from oil-based pesticides, pharmaceuticals, and industrial chemicals. Since the publication of Rachel Carson's *Silent Spring*, scientists have linked low-level exposure to these endocrine disrupters, which target hormone function, with declining male reproductive health, among other problems. In Denmark, for example, one of Europe's most highly energized nations, 30 percent of males show semen quality "in a subfertile range." Noted a 2006 study by Niels Skakkebæk and colleagues, "Our analysis suggests that we may, in fact, be seeing the signs of modern lifestyle influence on our reproductive capacity."

Meanwhile, billions of humans assisted by billions of energy slaves continue to work the globe like a cotton plantation. Today, the human population corrals 40 percent of the earth's plant-energy flow for food production. It consumes 35 percent of all biological productivity in the oceans. Humans have dammed, diverted, or monopolized 65 percent of the world's freshwater runoff. Approximately four-fifths of the earth's land surface and 98 percent of the area suitable for rice, wheat, or maize production has been shackled by the energies of humans. Some experts predict this petroleum plantation will extinguish half of the world's mammal species by 2050. The World Wildlife Fund estimates that "even with modest UN projections for population growth, consumption and climate change, by 2030 humanity will need the capacity of two Earths to absorb CO_2 waste let alone keep up with natural resource consumption." Other studies conducted by ecologists suggest the debt is much higher, due to the speed at which resources can be mined and their waste products thrown away. They conclude that if humans don't curb their consumption habits, we will need up to 2.8 to 5 Earths by 2050. That means the extinction of most nonhuman life forms.

The impact of human population on the natural world had reached a frenzied peak by 1980, the year business professor Julian Simon made a wager with scientist Paul Ehrlich, author of *The Population Bomb*. It may be one of the most famous bets of the last century. Ehrlich argued that exponential population growth was untenable and would lead to collapse and ruin. Sooner or later the real cost of raw materials would rise, and there would be shortages of food and water. Simon, a protégé of the oil-funded Cato Institute, countered that more people meant more ingenuity and that real costs would ultimately drop over time.

The two men chose five commodities to track: copper, tin, nickel, chrome, and tungsten. A decade later they reviewed the prices. The world's population had grown by 800 million by 1990, and the price of all five commodities had dropped. Ehrlich paid up, and the cornucopians rejoiced. They even claimed prosperity was infinite. But it was only Ehrlich's timing that was off.

Now that the era of cheap oil has ended, Simon would have grandly lost the bet. As *The Economist* reported in 2011, "an equally weighted portfolio of the five commodities is now higher in real terms than the average of their prices back in 1980." Had the list of commodities been broadened to include raw stuff such as phosphorus, potassium, and wheat, the scale of Simon's defeat would have been greater. Jeremy Grantham, a U.S. fund manager of much repute, reported in April 2011 that until 2002 all commodities except oil declined in price. After 2002 the price of everything else, from coal to cotton, began to rise. This commodity price surge was a paradigm shift, Grantham said: the era of abundance is over. The market is sending us "the Mother of all price signals . . . From now on, price pressure and shortages of resources will be a permanent feature of our lives," he warned. "The world is using up its natural resources at an alarming rate, and this has caused a permanent shift in their value. We all need to adjust our behavior to this new environment. It would help if we did it quickly." The Ehrlich-Simon wager, Grantham added, proved resolutely that humans are mortal and make short-term bets. Grantham, an optimist, thinks a saner world will require different energy and agricultural technologies, and probably a much smaller population. A slow, voluntary decline that reduces world population to perhaps as low as 1.5 billion might get the job done.

Demographers now talk about a third demographic transition: depopulation. Aging societies with falling fertility rates now perform the same function that famine, war, and epidemics once played in population control. Gary Peters, a U.S. geography professor, recently commented on the new model: "It assumes that the era of cheap fossil fuels has ended, no matter what effort is made to maintain it. That assumption is based on the fundamental concept that oil is both an extraordinary source of energy and finite in supply. In addition the model assumes that food production will decline worldwide as oil becomes more expensive. In turn that will lead to higher food prices, so feeding the poor will become more tenuous."

But François Cellier, a Swiss researcher on physical systems at ETH Zürich, doesn't think getting world population numbers back to normal will be easy even with dropping fertility rates. The average person on the planet now uses approximately 5.4 acres to make a living and provide for food, energy, clothing, and shelter. High-energy Americans occupy 24 acres, while low-energy folks in Madagascar get by on 1.2. If the world's land base were divided equally, everyone would be entitled to 4.4 acres. (With recent population figures, that share has gotten even smaller.) Currently, only Cuba occupies that spot. "If we wish to live in a sustainable fashion like the Cubans, we'll need to reduce our numbers by 20% to 5 billion people," says Cellier. "If we wish to all live like Americans, we shall need to decrease our numbers to roughly 1 billion people. Finally, if we decide to live as poorly as the people of Madagascar, then we can triple our numbers to 20 billion and live unhappily ever after."

Most models of world population growth show that the high consumption of resources and energy will lead to die-offs between 2030 and 2070. These culls could reduce the number of humans from 7 billion to 1 billion. But Cellier's work shows that the sooner societies stop using fossil fuels, the sooner this outcome will occur. Looked at from another angle, prolonged dependence on coal and petroleum is a one-way street to massive population collapse around 2040. (The Italian physicist Ugo Bardi calls this the "Seneca Effect": "Increases are of sluggish growth, but the way to ruin is rapid," as the philosopher wrote.) Other models based on dramatic reductions in fossil fuel use show major adjustments but not major die-offs. "The reason is that after the end of cheap oil the exponential growth pattern cannot be preserved any longer," Cellier explains. "The sooner we get out of the exponential growth pattern, the better we'll be off in the long run." Cellier proposes that people use fossil fuels sparingly and create a low-energy infrastructure.

Various groups have proposed targets for population reduction. England's Population Matters wants to whittle down the population of the United Kingdom from 61 million to 21 million. Australian environmentalists argue that the people "down under" should reduce their numbers by half, to 10 million. The U.S. organization Negative Population Growth suggests that the optimal number for that country is 150 million, a figure overshot fifty years ago. A one-child policy might achieve these goals over an eighty-year period, but a world populated by few children and many pensioners could ultimately invite new forms of eugenic engineering. "The latter half of the twentieth century may someday be seen not as ushering in the end of history, but as a transition out of modernity and into a new, prolonged, and culturally novel era of population shrinkage," notes Stanley Kurtz. "In any case, the social innovations of the modern world are still being tested, and the outcome is unresolved."

If humans fail to contain population growth, Mother Nature may well do the job. Vaclav Smil, the noted energy analyst, has said that "we are overdue for a pandemic," and he's probably correct. For more than a century, biologists have marveled at how viral epidemics often prune wild-animal numbers. These rambunctious plagues have winnowed wildlife population growth as effectively as have climate, predators, and changes in food supply. Animal ecologist Charles Elton watched viral outbreaks decimate rabbit, lemming, and sparrow populations with great fascination, noting that these disturbances were always associated with "over-crowding in the population." Great density in nature was usually followed by great scarcity, he added. These population fluctuations often occurred unexpectedly. In 1815, a mouse outbreak in Nova Scotia nearly undid the place. Hordes of mice invaded houses and farms. Swarms of cats, dogs, foxes, and martens did their best but made no dent in the plague. The mice devoured much of the hay and corn over a four-thousand-square-mile area. And then the mouse boom went bust. "They were seen crawling about slowly in a languid way, and then began to die in hundreds," wrote Elton. "Next season there was hardly a mouse in the whole district."

Frank Fenner predicted a similar fate for an overpopulation of humans a year before his death in 2010 at the age of ninety-five. The famed Australian immunologist spent much of his life studying viruses. He also helped to eradicate smallpox, one of the great population checks on humans. (It killed half a billion in the twentieth century alone.) Humans stood on the edge of extinction, said Fenner, due mainly to our untidy numbers, overconsumption, and climate change. He thought the event might take one hundred years. "The Aborigines [in Australia] showed that without science and the production of carbon dioxide and global warming, they could survive for 40,000 or 50,000 years," Fenner said in an interview with Niall Firth. "But the world can't. The human species is likely to go the same way as many of the species that we've seen disappear... It's an irreversible situation. I think it's too late. I try not to express that because people are trying to do something, but they keep putting it off. Mitigation would slow things down a bit, but there are too many people here already."

Aristotle once compared the stages of life—youth, maturity, and old age—to morning, noon, and night. Oil has created a world with more nights than mornings.

7

The Urban Fire

. . . . .

"The chief function of the city is to convert power into form, energy into culture, dead matter into the living symbols of art, biological reproduction into social creativity."

LEWIS MUMFORD, *The City in History*, 1961

OIL AND ITS companion, coal, concentrate people and power, and there is no greater manifestation of this uncomfortable truth than the explosion of megacities around the planet. More than half the world's population now lives in man-made urban centers often billed by planners as the architecture of progress. These complex mazes of roads and buildings occupy 3 percent of the earth's land mass but consume the majority of its energy. So earth's new master is an urban beast, and its slaves are the world's resources. In 1970, only three cities claimed populations exceeding 10 million, but by 2015, the world will boast twenty-two such metropolises. Like an ailing, aging financial system, these urban constructions seemed poised for a great reckoning.

Historically, cities lived on the surplus wealth generated by natural flows of energy: the fertility of the land; the direction of the wind; the course of a river; the shape of a bay; the muscle of their citizens; the diversity of a nearby forest. Within these limits, the medieval city rarely extended half a mile. But megacities recognize no defined borders. Fueled by an almost military array of energy slaves, the megacity consumes commodities from all over the world. It is not hard to tell the difference between cities built by hand (Venice, Barcelona, Havana) and those constructed by machines (Shanghai, Miami, Las Vegas). The former served people; the latter serve energy and its inanimate clients.

By any definition, the modern city has become a human feedlot, an energy hog, and a carbon bomb. The OECD estimates that cities now monopolize the majority of the world's energy flow. Cities currently command 82 percent of the world's natural gas consumption, 76 percent of its coal consumption, and 63 percent of its oil consumption. With this energy, just twenty-five of the globe's largest cities make more than half the world's wealth. It is a development as unnatural as confined feeding operations for animals.

For most of human history, the city held a modest place in the scheme of things. Medieval London was a trim affair. It housed 60,000 people on 700 acres. Two-legged transport kept cities small and contained. When it took 150,000 fertile acres to supply a largely vegetarian diet to 500,000 people, cities didn't grow much. Until the 1800s, less than 3 percent of the world's population lived in a city. Some ancient centers served as spiritual asylums; other well-walled enterprises offered security or military might. But when the world ran on solar power and human muscle, cities stood as unique, well-defined entities dependent on carefully managed energy surpluses. Cities connected power and authority to rural communities, and those communities reminded cities of the true source of their revenue. Good cities lived in respectful balance with the countryside. Industrial machinery destroyed that balance with cheap things, cheap transportation, and even cheaper standards, says agrarian philosopher Wendell Berry. "Rome destroyed the balance with slave labor; we have destroyed it with 'cheap' fossil fuel."

One of the first thinkers to grasp the ramifications of the Industrial Revolution for cities was Patrick Geddes. The Scottish botanist, later the father of town planning, recognized that the quick reengineering of cities was erasing the memory of previous standards for beauty and grace. The ugly new towns thrown up near coal mines closely resembled slag heaps, Geddes thought. These new hydrocarbon regions, which obliterated any sense of place, represented a radical departure in the evolution of cities. Geddes called the regions *conurbations*—groupings of people for "the machine and market order." In 1915, the urban critic concluded that British industrialists had erected, "out of all this exhaustion of the resources of Nature," a way of living for workers that was, "essentially, of Slum character." Smoggy conurbations dissipated energy to accumulate wealth in order to manufacture "growing infinitudes of mean streets, mean houses, mean back-yards," wrote Geddes. Geddes, who believed homes should have gardens and cities should have souls, recognized the primacy of energy in shaping this new urban creature. The Industrial Age "turned upon getting up coal almost anyhow, to get up steam almost anyhow, to run machinery almost anyhow, to produce cheap products to maintain too cheap people almost anyhow and these to get up more coal, more steam, more machinery, and more people, still almost anyhow and to call the result 'progress of wealth and population.'" Oil would only accelerate the proliferation of these conurbations.

Lewis Mumford, the great American journalist, reflected further on Geddes's sharp observations in the early 1960s. In *The City in History*, Mumford noted that agrarian societies powered by humans, horse, or sail logically supported diverse population centers. Coal and the steam locomotive did not. Oil concentrated people and machines even more. Some conurbations grew up near coal deposits; others mushroomed on top of oil fields or close to manufacturing. The cities built by cheap energy in Europe and North America typically boasted no center and stretched over the countryside for no purpose other than commerce. "The torrent of energy that was tapped from the coal beds ran downhill with the least possible improvement of the environment: the mill-villages, the factory agglomerations, were socially more crude than the feudal villages of the Middle Ages," wrote Mumford. These energy-rich groupings did not generate art or culture or even excellent science, he pointed out. "Whatever capital surplus they generated sped off to the wealthy to fuel vain personal luxuries or philanthropies in other cities."

With conurbations setting the new global standard for city-making, the United States became the standard-bearer. As the skyscrapers of New York, Chicago, and Los Angeles rose, so too did those cities' sprawling slums and windowless factories. (When the tourist G.K. Chesterton spied New York's great towers in the 1920s, he quipped that America was a Pharaoh "who built not so much a pyramid as a pagoda of pyramids.") In 1850, just 12 percent of Americans lived in cities; by 1910, 40 percent had become urban dwellers. Today, more than 80 percent of the U.S. populace lives in cities or their suburbs.

Congestion was a "phenomenon too obvious to need proof" in America's burgeoning cities, wrote Mumford. Wherever engineers built new transit lines, the facilities rapidly proved inadequate to the task. Declared the *New York Herald* in 1907, "As quickly as it becomes possible to ride about the city at an increased rate of speed, and with the opening up of new lines of transit, business increases, population is attracted to the city, and the problem is vastly complicated." In 1917, writing in *The Geographical Review*, engineer Sidney Reeve highlighted a truth that armies of megacity planners still debate: each repeated application of science and engineering to the problem simply made congestion worse.

Crowds gravitated to the greatest concentration of energy because that's where they could "make the best, the easiest, and the most certain living," Reeve reasoned. But these high-energy cities were so noisy, so polluted, and so crowded that they created a massive flight to the suburbs. The automobile allowed Americans to live in factory-made single detached homes in the middle of cornfields serviced by malls and mega-churches. Social critic James Howard Kunstler called the resulting suburban sprawl "the greatest misallocation of resources in the history of the world. We took all of our post-world war wealth—and actually quite a bit of the wealth that we had accumulated for decades before that—and we invested it" in a style of living totally dependent on cheap oil. "And to make matters worse, we didn't build it very well in the first place. So as it begins to decay it decays very rapidly and becomes a very unrewarding place to live in."

TODAY'S MEGACITIES—NOISIER, more polluted, and more congested than ever—are highly wasteful machines. They devour rivers of water, piles of food, and lakes of oil, disgorging mountains of garbage and clouds of dust and carbon. In 2000, the urbanist Herbert Girardet took a hard look at London, England, "the mother of megacities," and discovered a pathological metabolism. London's 7 million people converted resources and energy, largely from elsewhere, into 15 million tons of solid waste. The city also consumed the equivalent of 22 million tons of oil every year and thereby fouled the atmosphere with 66 million tons of carbon pollution. Although its inhabitants occupied a surface area of only 610 square miles, the total land area needed to grow food, provide timber, and balance the city's CO_2 production amounted to an area of 76,000 square miles. That's 125 times greater than London's geographic footprint and nearly three-quarters of Britain's land base. In other words, London monopolized the energy flows of three-quarters of the island. In real terms, explained Girardet, this appropriation stretched far beyond Britain's borders: you could find London's footprint in the cattle farms of Brazil, the wheat fields of Kansas, the fisheries of China, and the tea gardens of Assam. "We can argue that we no longer live in a civilization. We live in a mobilization—of people, resources and products," Girardet said.

The appetite of Asia's new megacities is equally gargantuan. Bangkok, for example, imports steel and copper from Sweden, porcelain from China, cars from Japan, fashion labels from Italy, and machines from the United States, then excretes most of these oil-soaked commodities into landfills, canals, streets, and garbage heaps. The waste stream of Bangkok's 6 million people has grown from 7,300 tons per day in 1995 to 9,600 in 2007. Forty percent of the garbage is food. Air and noise pollution exceed all acceptable limits, while water pollution has turned many canals rank and foul. Due to the overpumping of groundwater, many megacities are also sinking into the ground. Some parts of Bangkok, the "Venice of the East," drop as much as an inch a year. If everyone consumed as much paper, beef, and seafood as the residents of Hong Kong, the world would have to be twice its size.

The explosion of these megacities has created a class of consultants and academics who actively promote more of the same. These dedicated urbanists, who employ some of the world's worst prose, argue that monster cities generate wealth and are therefore better for global health. Piling people into skyscrapers, they add, also makes for greater energy efficiency. Economist Edward Glaeser predicts that the city will triumph as long as it allows like-minded individuals to gather to create wealth. In this vein, urban consultant Richard Florida talks effusively about the marvels of a "creative class," while sociologist John D. Kasarda predicts that only authoritarian cities armed with supersized airports will survive in the race to become bigger and richer. The Chinese government has proved his point: without debate it flattens villages, dams, streams and destroys landscapes to build mega-regional conurbations containing up to 41 million people, explaining that "democracy sacrifices efficiency."

For every problem created by globalization, the megacity can secure some sort of fix, chime the experts. Flooding Indian farms in Calcutta and Mumbai, they claim, will lessen the cost of delivering water and education to the masses by 30 to 50 percent. Cities offer a wide range of comforts, they boast, as though such qualities did not exist in rural communities. With a bit of tweaking, such as high-energy intensive "vertical farming," say the technocrats, megacities can even do a better job of feeding themselves. "Urbanization is an inexorable global force, powered by the potential for enormous economic benefits. We will only realize those benefits, however, if we learn to manage our rapidly growing cities effectively," goes one typical declaration, from the McKinsey Global Institute. The only limit to the size of cities and their pace of growth, says the consulting agency, might be the ability of business and politicians "to manage the increased complexity that comes with bigger city size."

By replacing soil and trees with concrete and asphalt, the modern city has created what academics less enamored of these conurbations call "urban heat islands." Pinpoint the densest concentration of traffic, roads, and high-rise towers in any temperate city, and there you'll find a major heat sink. (Asphalt roads soak up and radiate heat like solar panels.) When temperatures hit 84 degrees Fahrenheit in Sydney, Australia, the city's urban core records temperatures seven degrees higher. Since the capture of this heat bakes urban citizens, engineers are encouraged to burn more fossil fuels at power plants to air-condition overheated people—which in turn creates more GHG emissions and air pollution. In 2010, Paul Sheehan, a columnist with the *Sidney Herald*, fumed about the growing problem: "Modern culture is built around creating urban heat sinks, yet governments obsess less about this real-world, everyday problem than the more abstract problem of carbon pollution. Fixing the first problem would help ameliorate the second. But are there any grand plans for turning the web of our major city's blacktops into pale-surfaced roads? No. Any master plan for increasing the vegetation on footpaths and common areas? No. Any plans for retrofitting the kerb guttering and stormwater system so more water can soak into roadside green areas? No."

Big cities can even change local weather, causing more rain and thunderstorms. A combination of skyscrapers, waste heat, and air pollution, for example, makes Houston, Texas, a magnet for lightning strikes. Industrial petroleum sites in Louisiana suffer the same fate. Thanks to the stored energy released by concrete buildings and asphalt roads, temperatures in Bangkok, Shanghai, and Tokyo are steadily increasing. Tokyo has become such a heat engine that temperatures there have climbed 4 degrees Fahrenheit in the last century. A computer program called MEGAPOLI tracks the colossal impact on air quality and climate of six megacities: London, Paris, the Rhine-Ruhr region, the Po Valley, Moscow, and Istanbul. The pollution from Paris, for example, extends more than sixty miles outside the city.

Congestion in megacities is, of course, a fact of life. Weekday rush-hour automobile speeds often barely exceed six miles an hour. Cars in Bangkok, Delhi, Jakarta, Manila, and Seoul literally clog the streets, stifling economic activity while consuming extreme amounts of fuel. The complexity of the transportation networks in these conurbations can travel down two roads, warned Lewis Mumford years ago: "a breakdown of functions through neglect, or a financial breakdown through the increased expense of adequate service and repair." The U.S. urban infrastructure erected in haste by cheap oil now requires a $2 trillion upgrade solely to keep aging and eroding roads, ports, and air transportation networks safe.

Most megacities, in which a third of the inhabitants live in slums, already have little or no water security. The end of cheap oil means it will cost more to move, treat, and clean water. Mexico City, which grew from 2 million people to 21 million in just fifty years, has pumped its aquifers nearly dry. Each year the ground sinks another sixteen inches. In Buenos Aires, the Riachuelo River offers 12 million people water contaminated with heavy metals fifty times above legal levels. In dusty Nairobi, 60 percent of the city's 4 million people buy their drinking water from kiosks. Fouled rivers combined with climate change and the saltwater flooding of drained aquifers could destabilize Shanghai and its 23 million residents.

Cities and megacities fueled by hydrocarbons seem to follow some surprising biological rules. Luis Bettencourt and Geoffrey West, researchers at the Los Alamos National

Laboratory, combed through data sets on hundred of cities to figure out what makes them tick. They discovered that when it came to energy, large growing cities resemble large species of animals. An elephant might be ten thousand times bigger than a guinea pig, but it consumes only one thousand times as much energy. Big cities generally follow the same scaling rule. Whenever a city doubles in size, it requires an increase of only 85 percent in the number of streets, gas stations, stores, and homes. The 15 percent savings come from higher densities and better energy efficiencies. But as West and Bettencourt discovered, this equation has a dark side: crime, traffic, lung diseases, and general complexity all increase along the same lines. "What this tells you," says West, "is that you can't get the economic growth without a parallel growth in the spread of things we don't want." And not only that. Continuous adaptation, as opposed to stability, appears to be the daily bread of megacities. Without reliance on fossil fuels, West explains, "we broke away from the equations of biology, all of which are sublinear. Every other creature gets slower as it gets bigger because of the vast calories needed to keep moving. That's why the elephant plods along. But in cities the opposite happens. As cities get bigger, everything starts accelerating. There is no equivalent in nature. It would be like finding an elephant that's proportionally faster than a mouse."

West and Bettencourt herald the megacity as a foundation for creativity, power, and wealth. But the researchers also recognize that these behemoth's metabolisms can't be sustained without a constantly increasing supply of energy, water, and people. "The only thing that stops the superlinear equations is when we run out of something we need," West told the *New York Times*. "And so the growth slows down. If nothing else changes, the system will eventually collapse." In a 2010 paper published in *Nature*, the two scientists summed up the problem as they see it: "Cities are the crucible of human civilization, the drivers towards potential destruction, and the source of the solution to humanity's problems."

Nowhere in the world has the growth of cities advanced in a more warlike fashion than in China. Over the next twenty-five years, the country will build half of the world's new buildings, including some fifty thousand skyscrapers—the equivalent of ten New Yorks. China's bureaucrats even dream of creating "the endless city." *Guardian* reporter Jonathan Watts recently documented in his shocking book *When a Billion Chinese Jump,* that while Britain has 5 cities containing more than a million people, China can count 120. Most have names unfamiliar to Westerners: Suqian, Xinghua, Chongqing. It took China the burning of 55 million tons of coal each year and the addition of two new coal-fired plants per week to achieve this feat. Half of the freight carried by rail in China is coal. For this urbanization, 170,000 coal miners have sacrificed their lives; that's how many miners have died in mine explosions and collapses over the last decade. Since the 1980s, more than 400 million farmers and rural dwellers have flocked to China's cities. Chongqing, now a city of 28 million, grows by half a million people a year. A slave class of 100,000 porters armed with bamboo poles provides cheap energy for rich entrepreneurs in the mountain port. By 2020, China's cities will produce 400 million tons of waste annually—the amount the entire world vomited in 1997. Since the 1980s, its cities have destroyed 4.6 million square miles of land. The growth bears the tedious mark of standardization. "Many cities have a similar construction style," complained one bureaucrat.

Thomas Campanella, an American urban planner, compares the unprecedented scale of razing for China's newfangled conurbations to the wartime bombings of Dresden, Tokyo, Hiroshima, and Nagasaki: we've "never seen so much destroyed in order to build," he says. Rich agricultural lands the size of New England—44,000 square miles—have been cemented over. Speed and scale characterize the sprawling growth of Chinese cities. To Campanella they look like a "supernova, like an explosion across the landscape." A whole floor for an office building can be finished in three days, a high-speed rail system in only two years. Entire towns appear in the space of months. China, Campanella says, has broken all records for "the biggest bridges, the longest tunnels and the tallest buildings." It took thirteen years to build Heathrow Airport, but the Chinese-built Beijing Capital International Airport took just three. Quality, longevity, and safety don't appear to be part of China's city-making epic, however; shopping malls collapse and high-speed trains derail with headline-making regularity. According to Campanella, "Countless Chinese buildings, thrown up in haste, have already outlived their usefulness. The life span of architecture is measured in dog's years." A dozen office buildings built in Nanjing in the 1980s were already slated for demolition by 2006. "Binge building yields a high quotient of urban junk," says Campanella. In many respects, China has simply magnified all the mistakes of North America's oil-fueled cities.

China will soon also have the world's largest national highway system. Given that the country plans to build an additional 430 billion square feet of commercial and residential conurbations over the next two decades, McKinsey & Company estimates China will need a continent's worth of cement, minerals, and timber. Most of that material will come from Africa, where China has invested billions in infrastructure to extract the continent's wealth. "The endgame in that picture is not pretty," says Campanella. "So, worst-case scenario, China goes down the same oil-slicked path that we've been on." China could go green (it has the world's largest solar industry), but corruption and energy politics make that unlikely. Even if China were to start to build and deploy renewables on a grand scale, energy specialist Vaclav Smil doesn't think the output will come near to satisfying the extreme energy appetites of the nation's supersized cities. "China before long will have dozens of cities with five, eight million people," he points out. "How do you run a city like that on a wind turbine or a photovoltaic cell? How do you run modern megacities where most of the population would be housed in high-tower structures... on renewable energy sources?"

FEW CRITICS have written about the suburbanization of cities as eloquently as Lars Lerup. The Swedish architect and university professor lives in Houston, the world's energy capital. Like most modern cities, Houston is a chaotic testament to the freedom afforded by oil and the pursuit of individual pleasures. This suburban mutant, which Lerup calls "a conurbation of a third kind," occupies a million acres but has no zoning laws. Over one hundred years, its proud builders transformed a wet, hard-soiled prairie covered in oak trees into an urban conglomeration fragmented by speedways, parking lots, shopping centers, and lonely suburban estates. The city even filled its vital bayous with concrete sewers in an attempt to dominate the flow of water. Lerup likens the city's skyscrapers to pumpjacks: "The oil gusher is transfigured and petrified in the priapic tower—the emptying of the earth and the filling of the sky."

This massive reengineering of the landscape has created what Lerup calls a toxic environment. During ozone alerts, Houston's hospitals fill with children and the elderly. After a hurricane, the streets remained flooded for weeks "far beyond the assigned flood plains." Lerup foresees the day when devastating winds could knock out the city's power supply. That loss would take down the city's water treatment and supply system. The human-engineered sewage system could quickly become a delivery system for disease.

Each suburban city has its own hidden disaster scenario—some worse than others, says Lerup. "All such scenarios are characterized by the collusion of natural and artificial events. Los Angeles has its earthquakes, Phoenix its heat and drought, Houston and New Orleans their hurricanes, Mexico City its thermal inversions and Randstad, Holland, its inundations." Lerup doesn't think any form of social consciousness will change the makeup of megacities or of suburban maverick centers like Houston. Nor does he place much hope in increasingly aggressive technological innovations. In the end, he suspects, plagues, floods, earthquakes, and hurricanes will prevail. In the aftermath, cities will rebuild on a human scale. "We must again give room to our oceans, rivers and deltas," he writes. "Our destiny cannot be the further proliferation of cul-de-sacs."

Lerup may well be right. Many of the world's megacities, from Shanghai to Mumbai, are located on coastal plains near the ocean. By all estimates, climate change will eventually submerge parts of these cities. Rising sea levels will flood slums, poison freshwater aquifers, and force great migrations. Just a three-foot sea-level rise could cost Mumbai, a city of 25 million, $2,300 billion. Most urban sewage systems can't handle the extremely heavy rains that climate change now delivers. Storm surges could also flood and erode our conurbations in unpredictable ways. Most of these megacities won't be able to afford the cost of rebuilding or relocation.

The world's megacity boom has only one clear antidote, and that was prescribed by a small Austrian with a deft sense of humor way back in 1941. That's when Léopold Kohr, a radical economist and a friend of both George Orwell and Ernest

Hemingway, penned an engaging book about the pathologies of bigness called *The Breakdown of Nations*. Kohr's book, which was finally published in 1957, became the basis for E.F. Schumacher's *Small Is Beautiful*. Both books alienated people on the left and the right. "Whenever something is wrong, something is too big," wrote Kohr, who had fled to settle in the United States after Austria was annexed by Nazi Germany. "If the human body becomes diseased, it is as in cancer because a cell, or group of cells, has begun to outgrow its allotted narrow limits. And if the body of a people becomes diseased with the fever of aggression, brutality, collectivisms, or massive idiocy, it is not because it has fallen victim to bad leadership or mental derangement. It is because human beings, so charming as individuals or in small aggregations, have been welded into overconcentrated social units such as mobs, unions, cartels or great powers. That is when they begin to slide into uncontrollable catastrophe."

Unlike most contemporary urban thinkers, Kohr recognized that the mobilization of petroleum, electricity, and nuclear power hasn't changed the nature of our social difficulties; it has simply magnified them. "What matters is no longer war, but *big* war; not unemployment, but *massive* unemployment; not oppression, but the *magnitude* of oppression: not the poor, who Jesus said will always be with us, but the scandalous size of their multitudes."

Small city states, in contrast, which dominated the landscape of our forefathers, offer everything on a human scale. They encourage civic participation and are walkable. Small cities have produced many astute problem solvers, Kohr said, including Heraclitus, Plato, and Socrates. The Italian city-states enriched the world with Dante, Michelangelo, Raphael, Titian, and Tasso. Small German cities produced Goethe, Heine, Wagner, and Bach. Aristotle, the product of a small city, understood limits in a way many contemporary philosophers and physicists can't seem to: "To the size of state," Aristotle said, "there is a limit, as there is to other things, plants, animals, implements; for none of these retain their natural power when they are too large or too small, but they either wholly lose their nature, or are spoiled." The problem with big cities and big states, wrote Kohr, is ultimately the scale of their wastefulness: they don't generate energy—they absorb it.

To prove his case for small places, Kohr cited a novel 1946 report by the U.S. Senate. The report noted that conurbations, in which a few foreign firms employed a large percentage of the workforce, suffered extreme unemployment during the Great Depression. Small cities, blessed by a diversity of businesses, did not. Moreover, small cities boasted a larger middle and upper class, with genuine local loyalties. "Accordingly, these small-business cities had more civic enterprises, better co-operation with labour in civic affairs, and a better city to live in. Statistics were there to prove the point. The small-business cities had less than half the slums and a much lower infant death. They had more magazine subscribers, more private telephones and electric meters, more church members, and bigger libraries and parks."

According to Kohr, who taught in Puerto Rico from 1955 to 1973 then moved to Wales, endurable cities don't squander energy, and they employ properly scaled tools for building. "For pyramids, cathedrals, factories, roads are in the last analysis not built by money or machinery which is scarce even in the richest country considering that they can never get enough of it, but by hands which are ample even in the poorest communities, and represent the only alternative energy supply which can never be exhausted because everybody is born with it," he observed in his 1983 speech accepting the Right Livelihood Award. "But once again, for the intermediate technology of muscle power to be economical, the society served by it must be small, as I can see every day in my alternative little town of Aberystwyth in Wales, where I can achieve more by foot, which costs nothing, than by car, which costs a lot and with which I can do nothing at all except leave town. So, let us solve the one insoluble problem of our time, the high altitude disease of excessive size and uncontrollable proportions, by going back to the alternative to both right and left of a small-scale social environment with all its potential for global pluralist co-operative and largely unaffiliated selfsufficiency not by extending centralized control but by decontrolling locally centred and nourished communities, each built around a nuclear institution with a limited but strong and independent gravitational field of its own as it existed in the form of medieval monasteries."

Lewis Mumford's classic *The City in History* remains both a mirror and a forecast for petroleum's conurbations. The absolute power now wielded by megapoli confounds the city's original mission: the cultivation of right livelihood. Mumford, who wrote before petroleum had saturated every endeavor with the economics of productivity and efficiency, did not view cities as simply wealth creators. He saw them as moral compasses. By concentrating fossil-fuel energy at the expense of human energy and human concerns, megacities detonate the landscape. "Such power destroys the symbiotic co-operation of man with all other aspects of nature, and of men with other men. Living organisms can use only limited amount of energy. 'Too much' or 'too little' is equally fatal to organic existence. Organisms, societies, human persons, not least, cities, are delicate devices for regulating energy and putting it to the service of life."

8

The Economist's Delusion

. . . . .

"To act on the belief that we possess the knowledge and the power which enable us to shape the processes of society entirely to our liking, knowledge which in fact we do not possess, is likely to make us do much harm."

FRIEDRICH AUGUST VON HAYEK, Nobel Prize lecture, 1974

EVERY DAY nearly a million students learn that markets, money, and math, along with a good dose of greed, dominate economics. Thick textbooks and prestigiously degreed financial analysts tell the great story of minimizing costs and maximizing profits. Households purchase goods and services from big corporations, the story goes, and these corporations in turn buy or rent land, labor, and capital from individual households. Wages and profits animate the whole rational system, which flows like some miraculous river that widens exponentially every year without the need for rain or melting snows. Today's politicians genuflect to economists and their master computer models much the way kings once bent their knees to popes and scheming cardinals.

But this neoliberal model, whether used by monetarists (Big Markets Rule) or Keynesian economists (Big Government Rules), is a hydrocarbon comic book and an increasingly dark one as its central driver, petroleum, becomes more costly and difficult to mine. In insisting that labor, markets, and technology make the world go round, neoclassical economists have ignored the primary source of all wealth: energy. They have disregarded several thermodynamic laws and abused much math. They have also mistaken the creation and exchange of money for the production of real wealth. Oil has powered an unprecedented set of illusions: that exponential growth is normal; that self-interest is always rational; and that capital is disconnected from material resources. A once distinguished moral philosophy has degenerated into a bogus science whose experts offer predictions more inaccurate than daily weather forecasts. The ecologist Charles Hall puts it simply: "The abundance of oil allowed [economists] not to think about energy."

The work of economists has always reflected the intensity and quality of energy used by their societies. When solar-powered or slave-based agriculture ruled the day, economists mused largely about land, scarcity, and limits. When coal took over, "the dismal science" shifted its attention to the production of goods, "inexorable laws," and methods for distributing the surplus wealth generated by the new labor-saving machines. As oil greased the U.S. economy and then the global marketplace, economics morphed into a narcissistic endeavor that sought to turn an awkward social science into "the physics of society." In so doing, it made humans the master class. At the same time, it omitted energy from its models and dropped any mention of finite resources. It embraced markets and the flow of capital as the sole determinants of life. Even Marx and Engels, astute but often hyperbolic critics of capitalism, couldn't believe the proliferation of capital unleashed in the nineteenth century: "The bourgeoisie, during its rule of scarce one hundred years, has created more massive and more colossal productive forces than have all preceding generations together."

Freed from the constraints of human muscle and photosynthesis, economies that had once grown as slowly as trees now accelerated like trains. (They would later take off like jets.) European societies that burned hydrocarbons to power mechanical slaves grew exponentially, by 3 or 4 percent a year. The value of global economic production exploded from $700 billion in 1800 to $2.5 trillion by 1900. By 2000, it had reached a staggering $35 trillion. During the same period, average wages globally expanded from $600 to $6,000 a year. All in all, the world's economy in the late twentieth century was 120 times bigger than it had been in 1500.

This unprecedented eruption of consumption and spending correlated directly to increased flows of energy. The historian J.R. McNeill calculates that "the twentieth century used 10 times as much energy as in the thousand years before 1900 AD. In the 100 centuries between the dawn of agriculture and 1900, people used only about two-thirds as much energy as the citizens of the twentieth century." Yet mainstream economists still omit the supply of cheap energy from their forecasts to focus on money and technology.

In 1800, the world consumed just 440 million tons of oil equivalents in the form of coal, horses, human slaves, and wind. In 1990, it gobbled 33,000 million tons, largely in coal or oil. The correlation between oil spending and gross domestic product, the standard measurement for growth, looks like parallel lanes on a Los Angeles freeway. Since the 1950s, study after study has shown that when nations spend oil, their GDP rises; when oil prices rise, their GDP falls. The International Energy Agency calculates that every 1 percent increase in GDP requires a .3 percent increase in primary oil demand. The modern economy, says Swedish economist Kjell Aleklett, motors along: it simply won't run without oil. The implications, say Aleklett and the Uppsala Hydrocarbon Depletion Study Group, are obvious to everyone except modern economists: "We can now conclude that future growth in GDP must be dependent upon fuels other than oil if it is to continue as expected. This, in turn, defines the beginning of the end of the '*Oil Age*,' and society will have to seek other driving forces for future GDP growth."

The evidence of oil's dominion over economic matters is even more compelling in historical analyses. In 2011, University of California economist James Hamilton examined economic recessions in the United States since World War II. He found that ten of the eleven recessions (1960 was the exception) followed increases in the price of oil. Cheap oil fertilized economic growth, while expensive oil withered the crop. "The correlation between oil shocks and economic recessions appears to be too strong to be just a coincidence," he wrote. During the 1973 Arab oil embargo, an oil-price spike worth $5 billion in real market terms resulted in a $38 billion loss (a 2 percent drop in GDP) in the United States alone. "The dollar value of output lost in the recession exceeded the dollar value of the lost energy by an order of magnitude," says Hamilton.

Economic thinking was, at one time, more closely connected to energy realities. One of the world's first formal schools of economic thought belonged to the physiocrats. These eighteenth-century French agrarians and philosophers considered land to be the source of all wealth. François Quesnay, a physician who served the court of

Louis XV, believed that agriculture—the collection of solar energy—gave the economy its vitality. He correctly noted that in French society, surplus crops created economic growth, and he advocated for fewer tolls and taxes on farmers. Physiocrats also thought that a knowledge of natural and physical laws—including rainfall, growth, decay, and the turn of the seasons—should serve as the basis of economics. Quesnay, for example, paid attention in his calculations to how many horses and oxen a peasant employed, the size of his farm, how much feed he used. In other words, he catalogued energy inputs and their contribution to wealth. Quesnay and the physiocrats noticed that societies that observed natural limits tended to eat well. Those that failed eventually starved. For Quesnay, the aim of all economic effort was "to secure the greatest amount of pleasure with the least possible outlay."

Then along came a new crop of economists. In 1776, the Scottish moral philosopher Adam Smith published the world's first economic blockbuster for inanimate slave owners or manufacturers, *The Wealth of Nations,* in the midst of a coal boom. Smith's book asserted that economic progress depended on three essential qualities: the pursuit of self-interest, the division of labor, and the freedom of trade. Smith, a keen observer of economic fashion, preferred manufacturing to the export of "rude produce" from the land and considered industrial goods the new source of wealth. But unlike his later cult followers, Smith was no ideologue. He understood that there were limits to production, and he frequently acknowledged the fertile power of Nature in wealth creation. But he spoke nary a word about how the new machines forged and run by the heat of coal had allowed civilization to step beyond the boundaries of the solar world. He did, though, celebrate the consequence: "The great affair, we always find, is to get money." Physiocrats violently disagreed. Wrote Pierre Samuel du Pont de Nemours to Jean-Baptiste Say, a follower of Smith, "You have narrowed the scope of economics too much in treating it only as the science of wealth. It is *la science du droit naturel* applied, as it should be, to civilized society."

After Smith, laissez-faire economists overran the political landscape. David Ricardo—who began the unfortunate tradition of writing turgidly about economics—argued that countries should devote their labor and capital to whatever they did best. Given the prosperity of the British Empire and its newfangled industrialization, Ricardo and fellow thinkers like Jeremy Bentham and John Stuart Mill focused on the utilitarian generation of capital. Others focused on labor, with Karl Marx famously arguing that more capital should go to the working class. But the rapid increase in the flow of energy that made such debates possible escaped the classical economists' attention. For nearly one hundred years, capitalists and communists fought over what to do with the surpluses created by inanimate hydrocarbon slaves. They battled like brothers arguing over how to dispose of the family inheritance.

One of the first to denounce the folly of this ideological battle was Fred Cottrell. In 1955, the U.S. sociologist bravely named capitalism for what it truly is: a system that justifies the use of high-energy technologies. Wrote Cottrell, "Neither those who promoted Capitalism nor Marx and his colleagues recognized the way the shift from dependence on work done by food and muscle power to that done by fossil fuel and new converters had wiped out the hitherto necessary direct connection between the amount of product and human labor. Nothing in Marx's system would account for the fact that instead of increased poverty for the proletariat, which he predicted, the use of high-energy technology produced far more goods than could ever be produced by human labor alone, no matter how effectively it was deployed and motivated."

WHILE ECONOMISTS FEUDED, German physicist Rudolf Clausius discovered two vital thermodynamic principles. After studying the power produced by steam engines, Clausius noted in 1865 that energy couldn't be destroyed or created but remained constant in the scheme of things: the chemical energy released by the burning of coal to power a steam engine equaled the mechanical energy output combined with the heat produced. Humans could convert energy, but they couldn't make more than had been converted. Clausius's other discovery was just as key: energy lost its ability to do work during its conversion. More than two-thirds of the energy released by the firing of coal for steam, for example, simply went up in smoke or a haze of heat. This second law of thermodynamics showed that Nature sets limits: there could be no such thing as a perfect engine. Clausius called this inevitable loss of useful energy *entropy*. Entropy is partly the cost of doing work. But it also explains why people age and civilizations peak and corporations fall apart. A fallen apple does not return to the tree, and a car cannot run on its exhaust.

Clausius and his European peers immediately grasped the economic implications of his findings. Societies, the German engineer calculated, would eventually lose their heat (he talked grimly of "heat death"), and like many nineteenth-century scientists he worried about frittering away Nature's reserves. "We have found stocks of coal from old times. These we are now using and we behave just as a happy heir eating up a rich legacy."

"The black diamonds of England" were held in such high regard that classical economists viewed the coal smoke obscuring the nation's skies as nothing more than the cloud of progress. Coal animated the English with "inborn energy," and its abundance on the island was seen as gift of destiny. "Without coal we could hardly call ourselves a people," proclaimed a journal edited by Charles Dickens. But as coal's production declined, Stanley Jevons, a different kind of economic thinker, predicted that the energy of the British economy would dissipate too. After industry had dug up the richest veins, he wrote, miners would need to go deeper and would find less at greater costs. In 1865, Jevons warned that coal reserves would suffer exhaustion and "that the check to our progress must become perceptible considerably within a century from the present time." He also feared that British coal would lose its place in the world and be replaced by more abundant and cheaper sources from the Americas. The *Times* summed up his view in a review: "We may boast of the mental and bodily qualities which distinguish the Anglo-Saxon, and glory in the valorous achievement of our forefathers, we may delight in green pastures and golden corn-fields, and, like a certain noble lord, look forward with satisfaction to the day when our soot-begrimed Manchesters and Birminghams shall have crumbled into ruins and the plough shall pass over their sites; but this is certain, that without great wealth we should have remained in comparative obscurity; that without extensive manufacturing operations we should never have accumulated great wealth; and that without coal these operations would have been impossible."

The new intellectual fashion of money-seeking had no fiercer critic than John Ruskin. The popular painter, essayist, and social crusader—perhaps "the greatest Victorian bar Victoria"—called orthodox economics and its doctrine of enlightened self-interest a cheerful ratification of business as usual. A devout Christian, Ruskin argued that the laissez-faire pursuit of money through the use of mechanical slaves divorced human societies from their very foundations. He reminded economists that their new social science relied on preliminary sciences such as biology and geology, or what he called "the sound knowledge of living things." All wealth, Ruskin believed, depended on the inanimate flow of energy from Nature.

Ruskin saw the Industrial Revolution for what it was: the replacement of one form of servitude with another. In days of old, slaves had complained of compulsory work. Today, free men were slaves to machines, slaves to cutthroat competition and the joyless routine of mechanical labor. "But the modern Politico-Economic slave is a new and far more injured species," he wrote, "condemned to Compulsory *Idleness,* for fear he should spoil other people's trade; the beautifully logical condition of the national Theory of Economy in this matter being that, if you are a shoemaker, it is a law of Heaven that you must sell your goods under their price, in order to destroy the trade of other shoemakers; but if you are not a shoemaker, and are going shoeless and lame, it is a law of Heaven that you must not cut yourself a bit of cowhide, to put between your foot and the stones, because that would interfere with the total trade of shoemaking." Modern economists, Ruskin predicted, would extend their unnatural hold over society by disguising their discipline's superficial resemblance to science and presenting their ideas with a mathematical aridity and difficulty that, as botanist Patrick Geddes wrote in his study of Ruskin, "will at once keep off the public and impress them with profound reverence."

Another heretical voice, from Austria, was questioning the laws of supply and demand. Eduard Sacher, in *The*

Foundations of a Mechanics of Society, correctly traced all energy sources, from wind to coal, to their essential origin: solar radiation. Sacher regarded the sun as the world's first and only real capitalist. The Austrian economist painstakingly computed the energy available in coal, fishing, water power, draft animals, and even hunting. "The economic task of the available labour force," he wrote, "consists of winning from nature the greatest possible amount of energy." In his view, modern economics described primarily the appropriation of this surplus energy and the use of its interest, profit, and rent. In a world that saw global exchange as the source of wealth, Sacher's description of wind in the sails did not find an audience.

More to the popular taste were the ideas of Alfred Marshall, England's preeminent economist until his death in 1924. Marshall's textbook, *Principles of Economics*, ruled academic circles and ably spread the classical message. Labor and capital made wealth by answering to the call of supply and demand, wrote Marshall. "We are moving on at a rapid pace that grows quicker every year; and we cannot guess where it will stop... The whole history of man shows that his wants expand with the growth of his wealth and knowledge." Economic freedom was the primary objective; the entrepreneur who built "factories or steam-engines or houses, or rear[ed] slaves, reaps the benefit of all net services which they render so long as he keeps them for himself." To Marshall, the only energy that mattered was the drive to climb the social ladder.

ON THE OTHER side of the ocean, Frederick Soddy, a Nobel Prize–winning chemist, left his studies in nuclear science in Montreal during the 1920s to critically examine "what passes for economics." Wealth, he wrote, was an energy flow that could only be spent, not saved. Capital was energy embedded in certain products and so was subject to entropy: "If the supply of energy failed modern civilization would come to an end as abruptly as does the music of an organ deprived of wind." To expose the ignorance of modern economists, Soddy used the railway steam engine as an instructive metaphor. "In one sense or another, the credit for the achievement may be claimed by the so-called 'engine-driver,' the guard, the signalman, the manager, the capitalist, the share-holder,—or, again, by the scientific pioneers who discovered the nature of fire, by the inventors who harnessed it, by labour which built the railway and the train. The fact remains that all of them by their united efforts could not drive the train. The real engine-driver is the coal. So, in the present state of science, the answer to the question how men live, or how anything lives..., is with few and unimportant exceptions, 'By sunshine.'"

Economists, Soddy argued in *The Role of Money*, mistakenly viewed the acquisition and exchange of wealth as "tantamount to its creation." But physical energy in volumes far greater than that previously extracted from "the unwilling bodies of draught cattle and slaves" now made the production of wealth possible. The laws of thermodynamics applied to everything, Soddy wrote. Infinite wealth and even stable growth were illusions. Soddy viewed money as a limited good, governed and constrained by the available energy supply. Bankers did not care "for the interests of the community or the real rôle that money ought to perform." They even made debts subject to the laws of mathematics instead of those of physics, a mistake low-energy societies never made. (Medieval Europe banned usury for this very reason.) "Unlike wealth, which is subject to the laws of thermodynamics, debts do not rot with old age and are not consumed in the process of living," Soddy wrote in *Wealth, Virtual Wealth and Debt*. "On the contrary, they grow at so much per cent per annum, by the well-known mathematical laws of simple and compound interest... It is this underlying confusion between wealth and debt which has made such a tragedy of the scientific era."

Not surprisingly, mainstream economists regarded Soddy as a crank. The dismal trade continued to manufacture convenient new theories. Erich Zimmerman, a Texas economist much revered in the oil patch, proposed in 1951 that resources were no longer made by Nature but by men. "Resources are highly dynamic functional concepts; they *are* not, they *become*," Zimmerman wrote. He argued that oil was such a dynamic entity that it was best to foster a climate in which "resource-making forces thrive." Technology, capital, knowledge, and the strength of financial institutions made resource conservation "to protect the interest of future generations... unnecessary."

Following World War II, the lead on global economic thinking was seized by the United States, now a well-oiled petrostate. Economics became a justification for conspicuous consumption, and U.S. economists became celebrated gurus. In 1955, Victor Lebow described the role of a high-energy economy in the *Journal of Retailing*: "Our enormously productive economy demands that we make consumption our way of life, that we convert the buying and use of goods into rituals, that we seek our spiritual satisfactions, our ego satisfactions, in consumption. The measure of social status, of social acceptance, of prestige, is now to be found in our consumptive patterns. The very meaning and significance of our lives today expressed in consumptive terms. The greater the pressures upon the individual to conform to safe and accepted social standards, the more does he tend to express his aspirations and his individuality in terms of what he wears, drives, eats—his home, his car, his pattern of food serving, his hobbies.

"These commodities and services must be offered to the consumer with a special urgency. We require not only 'forced draft' consumption, but 'expensive' consumption as well. We need things consumed, burned up, worn out, replaced, and discarded at an ever increasing pace. We need to have people eat, drink, dress, ride, live, with ever more complicated and, therefore, constantly more expensive consumption. The home power tools and the whole 'do-it-yourself' movement are excellent examples of 'expensive' consumption."

In an economy stimulated by fossil fuels, explosive growth, and rapid technological change, U.S. economists imposed a seemingly scientific order on what was clearly an abnormal development in civilization. Paul Samuelson, perhaps the world's most influential modern economist, became as dominant in the field as Shell would in Nigerian politics. He added more theorems to his increasingly prestigious discipline, including one on the efficiency of markets. He also wedded the work of John Maynard Keynes to that of classical theorists to create a bastard child known as "neoclassical economics." Keynes had argued that it was correct for government to intervene and spend its way out of depressions and recessions, while others, such as Friedrich Hayek, preached caution. (To his later embarrassment, Samuelson also promoted the growth of bogus financial derivatives.)

With his textbook, *Economics*, the Nobel laureate standardized thinking about the economy the same way Rockefeller had standardized oil production. "I don't care who writes a nation's laws... if I can write its economics textbooks," Samuelson once boasted. The jaunty Indiana native said little about energy in his book, though he suspected that higher prices for goods might lead to lower living standards and less shopping.

The neoclassical economists generally served what they observed: increasing piles of money, generated by unprecedented energy spending. The libertarian University of Chicago economist Milton Friedman in particular championed free markets and less government. But like Samuelson, Friedman largely neglected the role of cheap energy in creating money flows. He emphasized the importance of stable growth for money supply but never defined what money truly represented. He never recognized that his monetary concerns were another symptom of entropy, of a decline in energy returns. Nor did he acknowledge how cheap energy built increasingly larger economic networks that nourished complexity instead of the freedom he touted.

Another U.S. economic titan preaching the gospel of endless growth was Robert Solow. For Solow, it wasn't capital and labor that now made the world go round, but technological innovations deployed by highly educated elites. After watching the automobile, aviation, and space industries take off in the 1950s, Solow calculated that 87 percent of economic growth could be attributed to technological progress. That cheap oil drove these new vehicles did not really concern him. In fact, Solow minimized the importance of finite resources. The Nobel Prize winner proposed that it was perfectly rational to drain the stock of any oil or mineral as long as one added "to the stock of reproducible capital," behavior few nations or companies ever practiced. In 1974, Solow considered the possibility that "the world can, in effect, get along without natural resources" because technology and money would always find a replacement. (He later amended that view: "It is of the essence that production cannot take place without some use of natural resources.")

By the 1970s, the basic tenets of neoclassical economics had taken on religious form. "Economic growth is the grand objective. It is the aim of economic policy as a whole," declared British economist Sir Roy Harrod. The neoclassicists believed that markets corrected themselves and operated independently of the natural world; that scarcity didn't exist, because the human mind would find alternatives; that technology (and not inanimate slaves) drove the future of economies; that individuals responded rationally to prices, always in their own self-interest; and that governments should let the marketplace get prices right. Sustaining the flow of money counted more than the flow of energy, biodiversity, or even human health. Economist Julian Simon took these libertarian reflections to a new level, reasoning that "the term 'finite' is not only inappropriate but downright misleading when applied to natural resources." Capitalism was a perpetual motion machine powered by people, Simon posited. The world's population explosion could correct resource exhaustion. As more scientists and technologists joined the economy, these new Einsteins would solve the problems created by unlimited economic growth; they would turn every scarcity and cost increase into a solemn benefit.

Many of these "pseudoscientific absurdities," as the philosopher Jacques Ellul has called them, are recycled today on MasterResource, an economics blog started by Robert L. Bradley Jr. and partly funded by ExxonMobil. Bradley, a former Enron associate and libertarian who in 2004 coauthored *Energy: The Master Resource*, claims that the world's material progress is "the result of advances in energy technology made by people living in freedom" and so will continue unerringly. The real enemies to growth, Bradley claims, are not doom-sounding depletionists but rather Big Government "statism" and environmental philosophies that set limits on drilling in parks and oceans and on public lands. "While the prices of individual fuels may rise," writes Bradley, "there is little reason to believe that energy *per se* will grow less abundant and more costly. The lesson of history is that in free societies individuals produce more energy than they consume." Resources, adds the ever-optimistic Bradley, "lie not in what can be seen but in what can be envisioned. They are limited only by the boundaries of our minds and by the physical universe."

Bradley does, however, acknowledge the importance of inanimate slaves. Thanks to hydrocarbons, the proportion of industrial work performed by human hands in the United States has fallen over the last hundred years from 90 percent to 8 percent. This blessed emancipation has given each American the fossil-fuel equivalent of about three hundred slaves, and Bradley predicts that the number of virtual slaves will only grow. "It is hard to overstate the significance of this trend. It means not just more creature comforts but a fundamental change in the human condition," he writes. "If we take the current population of the United States as being about 280 million people then the country as a whole has an equivalent of 84 billion" energy slaves.

Since energy consumption peaked in the Western world in the 1970s, Solow and a few other economists have expressed doubts about their profession's increasing reliance on mathematical models and formulas. In a 1982 letter to *Science*, Wassily Leontief expressed alarm at how insular his field was becoming: "Page after page of professional economic journals are filled with mathematical formulas leading the reader from sets of more or less plausible data but entirely arbitrary assumptions to precisely stated but irrelevant theoretical conclusions." Leontif urged historians, scientists, and even artists to add to the economic discourse. After the 2008 economic crash, Robert Solow reluctantly expressed his own misgivings about the disastrous dominance of economic modeling that had enriched a few at the expense of the many. In a statement to the House Committee on Science and Technology, he said, "They take it for granted that the whole economy can be thought about as if it were a single, consistent person or dynasty carrying out a rationally designed, long-term plan, occasionally disturbed by unexpected shocks, but adapting to them in a rational, consistent way. I do not think that this picture passes the smell test. The protagonists of this idea make a claim to respectability by asserting that it is founded on what we know about microeconomic behavior, but I think that this claim is generally phony. The advocates no doubt believe what they say, but they seem to have stopped sniffing or to have lost their sense of smell altogether."

Critics such as University of Laval economist Bernard Beaudreau have pointedly wondered why economists never talked about growth with engineers, who define the economic deity—referring back to growth as an increasing function of energy consumption and materials. To engineers, energy is the breakfast, lunch, and dinner that feeds capital and labor. All postwar growth in the United States, Japan, and Germany can be explained by massive infusions of industrial energy, argued Beaudreau in a 2001 academic paper. Yet "for unexplained reasons, the economics profession has ignored physics and thermodynamics in its work on growth, and more importantly... this omission has exacted a heavy toll."

The economics profession has also ignored the experience of the former Soviet Union. Neoclassical economists claim the Communist empire collapsed due to inefficiency, market planning, and Ronald Reagan's saber rattling. But that's not the story. Like the United States, Russia has always been well endowed with cheap oil. In the 1920s and 1930s, Communists began to use that petroleum to modernize a peasant economy. Like the United States, the country followed WWII with heavy investments in space technology and the military-industrial complex. With five-year plans financed by the proceeds of oil, the Soviet Union grew like modern-day China. Drunk on petroleum, it invested in Big Science and Big Education. Soviet technocrats poured money into nuclear power and renewable energy research, too. Dmitry Orlov, a caustic observer of the collapse, notes that the United States and the Soviet Union were merely high-energy antipodes: "Here are two 20th century superpowers, who wanted more or less the same things—things like technological progress, economic growth, full employment, and world domination—but they disagreed about the methods. And they obtained similar results—each had a good run, intimidated the whole planet, and kept the other scared. Each eventually went bankrupt."

In the end, oil pushed the Soviet Union over the edge. After oil prices collapsed in the 1980s, the state lost its central source of revenue. Oil production stagnated, then it peaked in 1988 as energy prices started to climb. Three years later, the empire fell apart in a sea of debt. The oil-assisted migration of 80 million grain workers to cities eventually added a bread crisis to the ferment. That emergency in turn forced a cashless government to borrow money to avert famine. Although glasnost, the Afghan war, and the Reagan doctrine all played minor roles in the collapse, the key change was that the Soviet machine ran out of fuel. Doug Reynolds, a professor of energy economics at the University of Alaska, spent two years in Kazakhstan before the collapse. The chain of events, he writes, was clearly energy driven. The Soviets, he says, "endured their first stagflationary shock in October of 1989, when their currency was devalued by 90%. Eventually, as the Soviet economy fell, each of the Soviet Republics from Lithuania to Kazakhstan left the Union. And during that whole time Soviet and post-Soviet oil production fell from a high of about 12 million barrels of oil produced per day to a low of about 7 million barrels a day, a 40% decline. So the real reason for the fall of the Soviet Union was an oil crisis. It was the third major oil crisis of the 20th century after the 1973 and 1979 oil crises, but you never hear of it."

After the collapse, Russia privatized its oil resources and introduced other reforms. These measures temporarily increased oil production, making the nation almost entirely dependent today on oil and gas revenue worth $200 billion a year. But Russia's renewed production of more extreme and expensive Arctic oil has restored neither stability nor the former level of public services. In fact, says Reynolds, Russia offers a global preview for high-energy societies faced with rising oil prices: "We can expect to see high unemployment and a collapse in the world economy. We can expect to see governments without any money to pay for things like health care, pensions, environmental problems, prisons, education or defense. We can expect to see infrastructure decay. We can even expect to see a decline in population. Finally, similar to the post Soviet Union, we can expect to see protests, political turmoil and revolution."

PERHAPS NO ONE has challenged the nostrums of neoclassical economics more cogently than Nicholas Georgescu-Roegen, a gruff Romanian who began his career as a student of Joseph Schumpeter. Georgescu-Roegen has called economics a "one-eyed discipline which sees only the market carried out by money." Like that of Frederick Soddy, much of Georgescu-Roegen's brilliant work, which appeared primarily in the

1970s and 1980s, was dismissed or marginalized by mainstream colleagues. But it remains strongly relevant today, and has earlier roots. In the 1950s, while he was standing on a bridge over one of Romania's biggest rivers, it struck Georgescu that human economics was an entropic affair: "I just kept gazing at its waters of dark chocolate color furiously running toward their final destination. Out of that simple panorama a definite thought came to me: There goes, I said to myself, our daily bread of tomorrows!" Modern high-energy economics, he wrote, reads like the tale of a human miner who inherited a bank and then dug a tunnel to rob the vault to speed up his ability to deplete its wealth.

Though he was an accomplished mathematician, Georgescu-Roegen renounced mathematical systems early in his career to describe economic phenomena, after Harvard economists predicted spectacular growth before the 1929 stock market crash. While studying his country's ever-shrinking Ploieşti oil fields and soil erosion in peasant farming communities, the Romanian recognized that economics was primarily about transforming highly valued natural resources (low entropy) into worthless piles of waste (high entropy). The long-term loss of fertile soil (a solar converter) and the depletion of oil fields were different forms of degradation.

Georgescu-Roegen's studies of peasant communities also convinced him that neither Marxist nor neoclassical economists understood human behavior. Economics was not a predictable story about individuals or super-individuals maximizing their gain. In fact, Georgescu found, peasant loyalty to tradition could produce economic results that outright violated the principle of product or utility maximization. Members of small agrarian communities often chose to act or consume for the good of the community. They wanted to make a living, not a killing. Pursuing maximum profit in an overpopulated economy didn't make sense if it put your friends out of work. To the Romanian economist's way of thinking, the discovery of fossil fuels had introduced the concept of self-interest to economics, not the other way around. Georgescu-Roegen had a knack for directness: "Only economists still put the cart before the horse by claiming that the growing turmoil of mankind can be eliminated if prices are right. The truth is that only if our values are right will prices also be so." (Decades later, the new field of behavioral economics pioneered by Daniel Kahneman and others would confirm that most people and most societies routinely make irrational economic decisions. Some people are motivated by shame or groupthink. Others will gamble recklessly with the aim of recovering losses. Moreover, people are far more altruistic and perhaps even more vindictive than quarreling neoclassicists: they will also deliberately lose money in order to punish wrongdoers.)

Georgescu-Roegen questioned the sophistry of economists who thought they lived in a limitless world. Inaccurate estimates about reserves of natural resources "does not prove the inexhaustibility of resources." And even resources that appear abundant may not be economically accessible. As for claims that scarcity could be dealt with by substituting "fewer resource-intensive goods and more of other things," Georgescu-Roegen just laughed: "He who does not have enough to eat cannot satisfy his hunger by wearing more shirts." Eternal growth was a "promise second in grandeur only to that of making everyone immortal." He was equally dismissive of economists who proposed a stationary economy. Such a miracle, he said, would sooner or later enter a crisis, "which will defeat its alleged purpose and nature." A person who thinks it possible to "draw a blueprint for the ecological salvation of the human species does not understand the nature of evolution or even of history—which is that of permanent struggle in continuously novel forms."

Yet the Romanian's peers continued to champion money, the portfolio theory, and other mathematical representations of economic Disneylands. But when the economic fraternity failed to predict the 2008 recession, fierce new critics emerged. Foremost among them was Nassim Nicholas Taleb, a former stock trader and business critic who argued that the Nobel Prize for economics should be revoked as a fraudulent charade. Taleb, author of *The Black Swan*, warned university students that current economic textbooks were as reliable and accurate as fifteenth-century medical treatises. In particular, he criticized prize-winning protégés of Solow and Samuelson such as Robert C. Merton, whose mathematical models justified risky financial derivatives.

Echoing critics like Soddy and Georgescu-Roegen, Taleb claimed that economic modeling failed because it omitted real risk and unpredictability. Neoclassical economists, who exhibit "a strange cohabitation of technical skills and absence of understanding that you find in idiot savants," have also ignored the basic teachings of Mother Nature. Earlier cultures not dependent on petroleum knew these lessons well. People feared debt and valued the importance of redundancy (having two bakers in a community), praised inefficiency (the source of quality), and appreciated robustness (waste not, want not).

According to Taleb, the inaccurate platitudes of Robert Solow and Paul Samuelson "have contributed massively to the construction of an error-prone society." A modern economist, adds Taleb, "would find it inefficient to carry two lungs and two kidneys... [His Mother Nature] would dispense with individual kidneys—since we do not need them all the time, it would be more 'efficient' if we sold ours and used a central kidney on a time-share basis. You could also lend your eyes at night, since you do not need them to dream."

Ecologists such as Charles Hall, a SUNY professor, have also called for the end of neoclassical economics. "Real economics is the study of how people transform nature to meet their needs," says Hall. "Neoclassical economics is inconsistent with the laws of thermodynamics." Author of the encyclopedic *Energy and the Wealth of Nations*, Hall proposes instead a new biophysical economics. Such a field would recognize that all wealth is created through the exploitation of natural resources and that economic policies should direct how energy is marshaled to exploit those resources. "Whenever per capita energy increases per capita wealth goes up," says Hall. He and Texas A&M researcher Jae-Young Ko conclude, "Since the availability of energy into the future is very iffy, then it would seem that the single most effective way for a nation to increase its per capita well-being is to reduce the number of capitas!"

Hall began studying economics because he was turned off by the modeling being done in the field of ecology. Whenever ecologists got away from musing about the behavior of real things (people, fish, trees), he noticed, they ended up with indefensible nonsense. He found exactly the same phenomenon in economics. In an email exchange he wrote, "The basic models just made no biophysical sense and there was a basic confusion of scientific rigor with mathematical rigor... I think that there is more intellectual creativity in attempting to figure out exactly how energy is the currency of ecosystems, including our economy, and how ecosystems respond to more and less available energy."

To demonstrate the superiority of biophysical accounting, Hall cites Costa Rica. Typical economists regard the country as a peaceful tropical paradise, a model of sustainability, and a great retirement investment for North American baby boomers. But that's not the picture Hall uncovered when he examined energy and resource use in the country. For starters, petroleum inputs to farming keep rising in Costa Rica, while yield per acre for most crops has dropped due to erosion, soil depletion, and fertilizer saturation. Since the nation is entirely dependent on imported fossil fuels, Costa Rica is extremely vulnerable to an increase in oil prices and eventual oil depletion. The country's continued population growth makes this problem more severe each year.

Hall argues, with Kent Klitgaard, that if the 5 to 10 percent of GDP that consists of energy costs were subtracted from the current global economy, most of the other 97 percent would disappear. "We are extremely lucky that we have to pay only the *extraction costs*, rather than the full-value *production, value-to-society* or *replacement* costs that Mother Nature might charge if there were mechanisms to do so," they write. "The full price would have to include the costs of natural capital depreciation, including both the fuel itself and the nature destroyed by its extraction, shipping and use, as well as the military costs of assuring resource availability. These we are hardly paying at present. If and when we run out of luck, and these costs come due, as will likely be the case, economics will become a whole new ball game in which the focus will return again to production and which will result in a new way of thinking about monetary and energy investments."

Robert Ayres, a physicist and economist at INSEAD (the "business school for the world"), has also begun to take the economic conversation in a new direction. Together with Benjamin Warr, Ayres has statistically shown that the highest phases of economic growth correspond with high and useful energy consumption. In the economies of the United States, Japan, and Europe, Ayres consistently found that "useful work does drive GDP growth." So the end of cheap oil will create volatile oil prices, stagnation, and even economic declines. And given peak oil and rising energy prices, the old notion that the next generation "will be a lot richer than we are" bears no relation to reality, says Ayres. Nor will the generations to come simply be able to buy their way out of the messes made by petroleum culture.

The standard economic notion that energy plays no role in the productivity of capital and labor, says Ayres, is "a very dangerous assumption and it's leading to potentially very risky advice to political leaders. For example, we're immediately faced with a possible inconsistency between the idea of taxing energy use, or putting a cap on carbon emissions—either of which will raise the price—while, at the same time, hoping that growth will not be affected."

Nearly forty years ago, Nicholas Georgescu-Roegen foresaw the day when inherited energy subsidies in the form of coal and oil would cost the world more energy than they returned: "We must emphasize that every Cadillac or every Zim [a large shipping container]—let alone any instrument of war—means fewer plowshares for some future generations, and implicitly, fewer future human beings." The world would eventually return to solar flows and experience a declining economy, the Romanian economist predicted. To avoid such a rude predicament, he offered a number of concrete recommendations, including the prohibition of weapons of mass destruction and a gradual decrease in population. Georgescu-Roegen believed that food should be produced solely through organic agriculture that employed people and animals instead of machines. He called for strict regulation of wasteful energy practices and an abandonment of "extravagant gadgetry." He thought we should make goods of better quality that were durable and repairable. Our leisure time should be devoted not to spending more energy but to making our surroundings more beautiful, our hearts more forgiving, and our minds more thoughtful. But the Romanian recognized that his ideas were perhaps impossible. Humans have a penchant to be human, and the energy crisis was really a crisis of the wisdom of the species. "Perhaps, the destiny of man is to have a short, but fiery, exciting, and extravagant life rather than a long, uneventful, and vegetative existence. Let other species—the amoebas, for example, which have no spiritual ambitions, inherit an earth still bathed in plenty of sunshine."

Today, the pseudoscience of economics, like science itself, has probably peaked. American economists Tyler Cowen and James Hamilton explain why with an economy of prose and thought rare among their peers. Cowen, an academic at George Mason University, is the author of a short best-selling book called *The Great Stagnation*. Hamilton, a California petroleum economist, penned a brief piece on exhaustible resources and economic growth that hardly anyone has read. But both men argue that the ideological assumptions undermining modern economics have begun to melt like glaciers.

Cowen's analysis barely mentions energy, but it does catalogue the symptoms of a thoroughly depleted enterprise: declining incomes, stagnating schools, complex government, and fewer innovations. "The average American family has been living with a growth slowdown for the last 40 years," writes Cowen. In 1972, U.S. domestic oil production and manufacturing peaked and the era of cheap oil quietly ended. In Cowen's analysis, the United States has already consumed the best resources and the lowest-hanging fruit. He offers just one typical American solution: "raise the social status of scientists" in an attempt to create better technologies. Cowen thinks that Americans should prepare "for the possibility that the growth slowdown could continue once the immediate recession passes."

James Hamilton, for his part, has documented that rising oil prices trigger recessions. Hamilton notes that cheap oil fueled the U.S. economy for most of the last century, and that the commodity was priced as though it was an inexhaustible resource. "By 1960," he explains, "the real price of oil had fallen to a level that was 1/3 its value in 1900. Over the same period, U.S. production of crude oil grew to become 55 times what it had been in 1900. On the other hand, the real price of oil rose 8-fold from 1970 to 2010, while U.S. production of oil fell by 43% over those same 40 years." Neither high prices nor technological improvements have altered this picture. Even with energy-intense and capital-demanding shale oil from North Dakota, U.S. oil production is 25 percent below 1970 levels, a time when oil was one-eighth of its current price. Given that most of the world's oil fields show signs of exhaustion, Hamilton writes, "there will need to be a shift in how the price of oil is determined." The depletion of easy oil combined with the acceleration of economies in China, Brazil, and India means "that the era in which oil is essentially regarded as an unlimited resource has ended."

As oil prices rise and become more volatile, citizens will spend more dollars buying fewer mechanical slaves. "Even a frictionless neoclassical model would conclude that the economic consequences of reduced energy use would have to be significant," writes Hamilton. In the process, the U.S. economy will slow down like a runner at the end of an impromptu race. "If the future decades look like the last 5 years, we are in for a rough time. Most economists view the economic growth of the last century and a half as being fueled by ongoing technological progress. Without question, that progress has been most impressive. But there may also have been an important component of luck in terms of finding and exploiting a resource that was extremely valuable and useful but ultimately finite and exhaustible. It is not clear how easy it will be to adapt to the end of that era of good fortune."

In a 2008 paper for the OECD, Swedish physicist Kjell Aleklett warned political leaders that the continent's economic stagnation and debt levels will grow worse as fossil-fuel liquids grow more expensive. (Like the United States, Europe now transfers $1 billion of its wealth every day to oil companies and petrostates.) "We have climbed high on the 'Oil Ladder' and yet we must descend one way or another. It may be too late for a gentle descent, but," Aleklett writes, "there may still be time to build a thick crash mat to cushion the fall."

9

Peak Science

. . . . .

"[The blame for the future plight of civilization] must rest on scientific men, equally with others, for being incapable of accepting the responsibility for the profound social upheavals which their own work primarily has brought about in human relationships."

FREDERICK SODDY, typescript, Bodleian Library, 1953

LONG BEFORE Frederick Soddy pinpointed the wobbly foundations of economics, the maverick thinker worried about the future of science. If civilization squandered its fossil fuels on conspicuous consumption, the father of nuclear fission said in 1910, it was possible that science, perhaps in the form of atomic research, might save the day. Yet he recognized an obstacle. With the discipline's "schoolboy successes" long past, the future of scientific endeavor lay not so much in discovering new things as in providing more consumptive services. Because the availability of energy slaves now drove innovation, Soddy forecast that science, a "fair-weather friend," would peak along with energy consumption.

To Soddy's mind, most academics and bureaucrats read science all wrong. Science "has been generally misunderstood as creating the wealth that followed the application of knowledge," he wrote. Instead, Soddy reckoned, science was just one of the lucky inheritors of the surplus energy flows generated by fossil fuels. If atomic energy failed to produce "private suns" for citizens, he said, then science, a child of hydrocarbons, might have to wrestle with a more violent change in social and political habits than that wrought by the Industrial Revolution. "Curious persons in cloisteral seclusion are experimenting with new sources of energy, which, if ever harnessed, would make coal and oil as useless as oars and sails," he lectured. "If they fail in their quest, or are too late, so that coal and oil, everywhere sought for, are no longer found, and the only hope of men lay in their time-honored traps to catch the sunlight, who doubts that galley-slaves and helots would reappear in the world once more?" Fossil fuels have energized science the way gasoline propels a car. For decades, chemists and physicists have pointed out that only about 14 to 25 percent of the fuel put into a vehicle actually gets used to move the transportation slave. The rest is lost in heat or squandered on distractions like heated seats. Powered by hydrocarbons, science performs the same way: it sometimes propels society down new paths with genuine discoveries, but most of the time it engages in elaborate forms of heat dissipation. Like the rest of the world, science has used fossil fuels to scale up and accelerate without much thought to the consequences. But without high energy flows, there wouldn't be much science at all.

The field's rapid ascension after 1820 confirms that science was, like abolition and massive city-making, an emission of concentrated hydrocarbon spending. Prior to the Industrial Revolution, the study of how the natural world worked was the preserve of a few curious scholars and philosophers. Solar- or human-powered science moved backwards and forwards, frequently stagnating. A remarkable calculator for measuring the motion of the planets, for example, appeared in Greece in 80 BC, only to disappear. Societies dependent on "the ancient wisdom of those who lived directly by sunlight," said Soddy, could not afford to devote much energy to scientific pursuits. There just wasn't enough surplus wealth to pay the bills. But thanks to coal, science took off like a steam train.

The heady phenomenon was driven by brilliant men and women from all walks of life, including Alfred Wallace, Charles Darwin, and Marie Curie; few were formally educated as scientists. Tellingly, it was the countries with the most coal and mechanical slaves, England and Germany, that dominated nineteenth-century science. Baron Justus von Liebig, the father of chemistry, often remarked that civilization was all about obtaining the greatest effect with the smallest expenditure of power; his own age was marked by an "extraordinary superiority of power." Science, wrote Liebig, not only "causes machinery to do work formerly done by slaves" but had "established a more just proportion between the forces of external nature" and humans.

While coal ruled global energy demand, scientists generally held that the world was finite, knowable, and subject to precise natural laws. ("Each year science increases by so many millions of horse-power its patient armies of inanimate slaves," lectured Frederick Soddy in 1918.) But then came petroleum and an even greater surplus of wealth. This volcanic burst, erupting first in the United States, transformed science from the untidy pursuit of truth to an efficient cadre of technicians intent on human improvement and progress.

It also made everything more complicated, note New Mexico anthropologist Joseph Tainter and Texas engineer Tadeusz Patzek. "We use more energy in more complex applications, and then need more complexity to manage the increased energy flows from society," they write. That has been the story of hydrocarbon-fueled science for nearly 150 years.

The speed and scale of scientific advances prompted soul-searching from the beginning, at least among some scientists. After startling innovations built oil-powered weaponry that left millions dead on the WWI battlefields of Europe, many expressed doubt and dread. Soddy was one of the first to bitterly reflect on the state of science and the "accelerating progress of the spendthrift to destruction." Most scientists found themselves alternately disturbed and seduced by their new role as masters of societal transformation. Alexis Carrel, the Nobel Prize–winning organ transplant pioneer, embodied the best and the worst of this scientific ambivalence. The French-born surgeon spent much of his career at the Rockefeller Institute for Medical Research trying to increase human longevity. He even kept chicken-heart tissue alive for thirty-four years. At the same time, Carrel criticized how science had standardized and degraded cities and factories through the careless deployment of mechanical slaves: "What is the good of increasing the comfort, the luxury, the beauty, the size, and the complications of our civilization, if our weakness prevents us from guiding it to our best advantage?" he wrote in *Man the Unknown* in 1935. Yet a few pages later he was famously proposing that murderers, thieves, and child kidnappers "should be humanely and economically disposed of in small euthanasic institutions supplied with proper gases." With so much energy on the loose, he concluded, only a small group of superior scientific individuals could properly direct "the ultimate purpose of civilization."

U.S. petroleum production reinjected science with a combustion-friendly morality and created what French philosopher Jacques Ellul called the American "scientific technopolis." If coal gave the world mechanized thought, then oil gave birth to technique, a powerful, practical science concerned with the mastery of chaos and instability. Oil and its related surpluses gave the world better weapons, a space industry, medical engineering, and vast information systems. U.S. scientists supported by the largesse of oil served the pursuit of happiness, consumer goods, and longer lives. Their chemicals and gadgets would not only enrich life, they promised, but tear down borders and create multiple Gardens of Eden. Sociologists, economists, and psychologists joined the race to be "scientific."

One of those to notice something amiss in this brave new world was the novelist and science writer Aldous Huxley. In his 1946 book *Science, Liberty and Peace*, Huxley offered a short, acerbic critique of high-energy society. He also pinpointed its greatest hazard: the concentration of power into fewer and fewer hands. In fact, Huxley warned, the advance of science had strengthened the progressive centralization of power in both big governments and big corporations. While supposedly disinterested experts equipped companies or the state with bigger machines for mass production or mass killing, they ignored communities and individuals interested in more robust or low-tech ways of doing things. Even individuals with specialized technical educations were not immune to deceitful propaganda, Huxley noted, and were capable "of the most dangerously irrational prejudice." Scientists, he added, had flocked in great numbers to work under the Nazis, Japanese militarists, and the Soviets.

Huxley, a keen observer of California's high-octane oil culture, argued that the concentration enabled by highly energized science had but one antidote: a popular decentralization movement that could develop local food and energy sources. Oil was a most "undesirable fuel from the political point of view" because it concentrated power in a few nations: "The world's strictly localized sources of petroleum, and the current jockeying for position in the Middle East, where all the surviving great powers have staked out claims to Persian, Mesopotamian and Arabian oil, bodes ill for the future." Huxley proposed that scientists take a sort of Hippocratic oath: "I pledge myself that I will use my knowledge for the good of humanity and against the destructive forces of the world and the ruthless intent of men."

Derek de Solla Price, a physicist and famous science historian, came up with some Huxley-like conclusions himself in the 1960s. Science not only mirrored exponential growth in the global economy and the population, Price argued, but required bigger institutions to manage its complexity. To accommodate the higher costs of increased energy expenditure in science (more lab workers), the profession had grown, with more institutions, national facilities, and global networks. Big Science not only reflected the concentrating tendencies of high-energy society but was death to resilience. Little Science, in the form of small independent labs, did not cost as much and worked agilely. It was also more responsive to community needs. Little Science could move fast and adapt to change. Big Science, however, worked like Big Oil. With its imponderable bureaucracies and its allegiances to Big Money and Big Government, Big Science often crushed innovative ideas and ignored simple solutions. Price didn't think the rapid multiplication of knowledge could continue indefinitely, either. A ceiling would be reached, at which point progress would become chaotic, a new innovation might arise, or the research would die out.

Price calculated in 1963 that 80 to 90 percent of all scientists who had ever lived now worked in facilities or institutions in Europe, North America, and Japan. "Alternatively, any young scientist, starting now and looking back at the end of his career upon a normal life span, will find that 80 to 90 percent of all scientific work achieved by the end of the period will have taken place before his very eyes, and that only 10 to 20 percent will antedate his experience." Price did not consider exponential growth in the field any more sustainable than exponential population growth. "It is clear that we cannot go up another two orders of magnitude as we have climbed the last five. If we did, we should have two scientists for every man, woman, child, and dog in the population, and we should spend on them twice as much money as we had. Scientific doomsday is therefore less than a century distant."

While Price made his calculations about the limits of Big Science, Jacques Ellul, one of Europe's greatest social critics, defined its dynamic technocratic character. (Ellul, a fiercely independent academic who wrote more than fifty books before he died in 1994, remains the world's foremost debunker of technological society.) For starters, Ellul recognized that technique, like the fossil fuels that energized it, was ambivalent by nature. It wasn't good, bad, or neutral, but simply "the driving force in all decision making." The discovery of penicillin, he argued, did not compensate for the devastation of Hiroshima by nuclear technique or the destruction of the Niger Delta by scientific oil-drilling practices. Moreover, technique's beneficial impacts could not be divorced from its destructive ones. Unforeseen consequences and side effects would multiply as technicians created novel and more powerful gadgets. Today, the same technique that makes it easy to book an airplane flight also makes it possible for banks, insurance companies, politicians, and charlatans to contact or monitor people twenty-four hours a day. Now that petroleum drilling technology can penetrate the earth and the ocean to depths of more than two miles, consumers shouldn't be surprised when such extreme behavior triggers earthquakes or contaminates groundwater or when highly complex facilities such as bitumen upgraders or offshore drilling rigs fail.

Fossil fuels and atomic energy had created a scientific world resembling a bright lamp around which people fluttered like moths, Ellul believed. He feared that science's growing complexity and increasing demands on ordinary people would make it more difficult for them to discern "what is necessary and what has meaning." Until digital machines stormed onto the world scene, for example, most people thought of world history in terms of changes in energy mobilization: from animals to slaves to fossil fuels and finally to atomic energy. The bright-screened computer and its related gadgets created the illusion that the world was now propelled by megabytes of information. "A new model of society was emerging," wrote Ellul. "The computerized society, the society of networks... supposedly liberates us from the constraints of energy, reversing entropy!"

Yet the multiplication of digital hardware, which at first appeared a marvel of energy conservation, was a complete energy bluff. For nearly one hundred years, industry had manufactured general goods by burning one pound of fossil fuels for every pound of plastic or metal product. A typical car, for example, required less fuel to build than it consumed during its lifetime on the road. The digital revolution turned this energy equation upside down. A laptop requires 26.5 pounds of oil for every pound of computer. Given that most laptops don't last more than three years, the majority of energy consumed in a computer takes place during its construction in Asian factories operating under slave-like conditions.

Technological obsolescence may represent the greatest oil spill the world has ever seen.

Computers run on microchips, another highly energy-intensive process. Most machine injection processing and casting takes enough energy per 2 pounds of product to run a flat-screen television for 1 to 10 hours. MIT researcher Timothy Gutowski recently calculated that semiconductor manufacturing demands enough energy to run that same flat-screen TV anywhere from 42 days to 114 years. In other words, the computer industry requires 800 pounds of fossil fuels to make 1 pound of microchips. "The seemingly extravagant use of materials and energy resources by many newer manufacturing processes is alarming and needs to be addressed alongside claims of improved sustainability from products manufactured by these means," say Gutowski and his colleagues. The world now types and clicks away on 1 billion personal computers and 3 billion cellphones.

The knowledge to be gained through computer use was another illusion, Jacques Ellul believed. We may lack good-quality food, wrote Ellul, but we will never lack "a superabundance of the empty nourishment of information." Although the world has increasingly lost access to forms of energy that didn't require the sacrifice of crops, forests, and groundwater, people could still stare at their screens "in the direction of nonsense."

And computers reflected but one aspect of Big Science and its wild energy appetite. Like Huxley, Ellul believed that the new aristocratic class of technicians held too much power. Their fixation with yield, speed, and efficiency had made the world as fragile as a Formula One racing car. The faster Big Science evolved, the more brittle the entire network became. "Vulnerability goes hand in hand with greater uncertainty," wrote Ellul, but the experts could not see fragility.

"Technical progress does not know where it is going. This is why it is unpredictable, and why it produces in society a general unpredictability."

The gaps in knowledge shown by modern scientists also stunned Ellul. "When these technocrats talk about democracy, ecology, culture, the Third World, or politics, they're touchingly simplistic and annoying ignorant," he noted. For every problem or challenge, they offered only one solution: more technology. (No one has employed this refrain more religiously than the oil and gas industry, which probably spends the least of any industry on innovation or technology. "Technology and innovation will enable more development of the oil sands," promises a lobby group. "Using technology to achieve sustainability" goes another advertising slogan.)

Ellul argued that the technopolis was subject to diminishing returns, if not entropy. His prediction is borne out today. Each new extreme oil technique, from oil sands to oil shale, seems to cost more, takes longer to develop, and typically exceeds predicted costs, only to deliver less volume and even fewer energy returns. "We cannot absorb indefinitely the repeated shocks of overwhelming techniques," wrote Ellul. "The more novel and powerful a technique is, the more it disturbs the world, and the more, in doing so, it contributes to the fragility of the technical system." Moreover, he said, "the growth of waste, of technical aberrations, of adverse effects, of slowness in the transition from innovation to application, and of subtracted values reflects the global reach of technique and leads inevitably to a lowering of productivity." Although Ellul believed that certain innovations could temporarily arrest declines or create the illusion of unending progress, he felt that the second law of thermodynamics applied equally to economics and technical progress. Even the nova-like explosion of specialization in science, he reasoned, represented a form of disintegration and a growing state of disorder. "There is no progress that is ever definitive, no progress that is only progress, no progress without a shadow. All progress runs the risk of declining. There is a double play of progress and regress. The 19th century ignored the shadow of industrial development and we today basically ignore the shadow of technical progress."

The evidence now largely supports Ellul, a man known for his foresight. Diminishing returns in science are quantifiable and can be seen most obviously in the world's science leader, the United States. When the National Science Foundation looked at publishing trends in the field in 2010, it found a downward spiral despite increases in funding. The report calculated that "the same resources that produced 100 publications in 2001 would have produced 129 publications in 1990." Its authors found the trend disturbing but offered no real explanation. To young science PhDs, the phenomenon expresses itself in low wages combined with a sixty- to eighty-hour work week. The Italian physicist Ugo Bardi observed that his colleagues now face "reduced budgets, more paperwork and the sensation [of] running a rat race." Even the Task Force on American Innovation acknowledges that government investment in science agencies peaked in 2004 and has been declining on average ever since. The nation that built a scientific technopolis cannot maintain it.

Diminishing returns can also be found in dramatic increases in fraudulent science. False tales about cold fusion and plutonium poisoning have been replaced with startling fictions about stem-cell research and snake-oil pharmaceuticals. To many observers, science fraud appears to be growing even faster than financial fraud. The number of retractions for fraud, mistakes, or plagiarism in scientific journals increased fifteenfold between 2001 and 2010. As one medical researcher scientifically noted, "Levels of misconduct appear to be higher than in the past." In China, the world's leading energy consumer, many scientists appear to specialize in science fiction. Most of the papers in the nation's five thousand science journals are largely published for show and go unread. Over two years beginning in 2008, 30 percent of the submissions to the *Journal of Zhejiang University-Science* were copied. Chinese scientists recently claimed to have discovered nine new turtle species that subsequently proved to be well known and purportedly invented a brand-new microchip that didn't exist. Critics of scientific fraud in China are routinely attacked, beaten, or abused. Around the globe, fraud in science has become so mainstream that ethical lapses support an entire website, *Retraction Watch.*

In many respects, the decline of science reflects diminishing returns from health investments to prolong human life. Thanks to antibiotics, clean water, and fossil fuels, U.S. life expectancy grew from forty-seven to sixty-eight years between 1900 and 1950. During that time, industry and government spent less than $300 million a year on health research. Between 1950 and today, U.S. life expectancy gained another nine years, to an impressive seventy-seven. In 2005 alone the United States spent nearly $30 billion on health and science research, while pharmaceutical companies deployed another $30 billion. Concludes Richard Epstein, a Chicago law professor, "No one can plausibly claim the material improvement in human life and enjoyment from these post-1950 changes matches in its impact the aggregate improvements made with much less investment in the period between 1900 and 1950. The second half of the twentieth century had no health miracle to match the computer or the Internet. The first half of the twenty-first century promises to be no different." Epstein identifies government as the key obstacle and recommends massive deregulation of the drug industry. When an industry can grow only by reducing government oversight, a peak indeed has been reached.

In 2005, Jonathan Huebner, a physicist at the Naval Air Warfare Center in China Lake, California, examined 7,200 innovations since the Dark Ages, relative to population. Contrary to economic claims that more people equals more brainy ideas, Huebner did not find an accelerating curve as global population increased. In fact, major technological advances per billion people peaked in 1873 and then declined. The number fell with higher rates of economic growth. It even fell "with higher levels of education, major advances in science and the invention of the computer." All in all, "it was harder for the average person to develop new technology in the 20th century than in the 19th." Most energy consumers don't realize that almost every innovation they use in their daily lives, including electricity, the steam engine, and the internal combustion engine, had been invented by 1873. By 2024, Huebner predicts, the rate of innovation (roughly three events per year) may decline to the level recorded during the Dark Ages. He suggests that the decline defines either the economic limits of technology or the limits of the human brain.

A similar peak can be found in almost every scientific endeavor. Much of the hundreds of billions now spent on medical research pours into genetic exploration. Scientists can map the genomes of various diseases with megabytes of data that require computer Leviathans. Yet these industrial forays have yielded few breakthroughs. Thirty years ago, a fraction of those research-dollar levels resulted in a number of critical innovations, from open-heart surgery to dialysis treatment. Now, with more money fertilizing more labs, fewer and fewer discoveries of note have been made. One scientist has even compared reductionist inquiries in genome mapping to "a mountain that "has laboured and brought forth not much more than a mouse." The more politic *Science* magazine complained that "the floodgates of understanding" have not yet opened. In one recent case, scientists mapped the genome of the bacteria responsible for the Black Death. Yet this bit of cleverness told them nothing new. The bacillus has not changed in several hundred years, which means that its formidable killing properties were due to a constellation of social and ecological events—a fact well known among scientists nearly one hundred years ago.

Some scientists now believe that scientific progress is subject to the same finite limits as oil itself. The scientific philosopher Nicholas Rescher, for example, contends that science will never be able to answer all the great questions. What is knowable will be limited by economic resources and energy. "Nature always has hidden reserves of power," Rescher says. Scientific progress will slow when the costs of technological tools to increase velocities, temperatures, and frequencies rise, as well as with the competing demands for scarce resources. The rising complexity of social problems will also ensure diminishing returns. Rescher doesn't foresee an end to progress, just a pronounced deceleration. "Once all of the findings at a given state-of-the-art level of investigative technology have been realized, one must move to a more expensive level... In natural science we are involved in a technological arms race: with every 'victory over nature' the difficulty of achieving the breakthroughs which lie ahead is increased."

No technology illustrates the confused state of peak science better than carbon capture and storage (CCS). Science academies in thirteen high-energy countries have dubbed CCS a "top priority" and a "vital component" in the fight against climate change. Canada and Norway, both oil-exporting states, hail CCS technology as imperative and have invested billions of taxpayers' dollars in trials and research. So too have the world's most powerful oil and gas companies. Like all bad ideas, CCS sounds attractive. The technology would collect emissions from fossil-fuel power stations, compress and liquefy them, and then inject the pollution into depleted gas or oil reservoirs or salt aquifers deep in the earth. Industry particularly likes the idea because it owns the transportation and injection technologies for what Shell calls "the only technology available to mitigate emissions from large-scale fossil fuel use."

But there is an energy paradox with CCS. Any coal-fired power plant, oil refinery, or bitumen upgrader equipped to handle CCS will pay a huge energy penalty: it will have to burn anywhere from 25 to 32 percent *more* fossil fuels. Stripping and compressing CO_2 is not cheap. This form of energy cannibalism also requires nearly a third more water and a third more chemicals. And the penalty doesn't include the energy costs of retrofitting these plants. Moreover, to bury just 20 percent of the world's emissions, calculates energy expert Vaclav Smil, we would have to create an entirely new worldwide absorption-gathering/compression-transportation/storage industry. That industry's annual throughput would have to be about 70 percent larger than the annual volume now handled by the entire global crude oil industry, whose immense infrastructure of wells, pipelines, compressor stations, and storage systems took generations and about $60 trillion to build. Monitoring CO_2 cemeteries would require more energy and public funds for thousands of years.

CCS represents the classic thinking of Big Science: it offers an energy solution that requires even more energy in an already high-energy society. Independent scientists regard it as thermodynamic travesty and bad engineering. Why manage emissions when you can simply reduce them? asks Vaclav

Smil. "Carbon sequestration," he explains, "is irresponsibly portrayed as an imminently useful large-scale option for solving the challenge." Many low-tech alternatives exist. These, he says, include banning automobiles in urban cores, raising fuel prices, imposing carbon taxes, improving public transport, building renewable energy projects, and protecting tropical forests. All would reduce emissions substantially with less energy, lower costs, and fewer scientists.

Just as ancient Rome supported slavery, Big Science champions tools that strengthen its energy and power monopolies. Gregory Unruh, a U.S. business analyst, correctly calls CCS another "carbon lock-in," business as usual for the "techno-institutional complex" of fossil fuels. CCS would strengthen the status quo, theoretically extending the fossil-fuel era by a few hundred years. Unlike renewable-energy projects, it would preserve the existing investments of multinational oil and gas companies in technology, know-how, and capital. The burial scheme could reduce some emissions, but it would also extend the period of CO_2 emissions. And by draining government coffers (CCS can't proceed without billion-dollar subsidies), the technology would not only delay more durable and cost-effective measures but divert resources away from low-tech solutions.

In an astute essay in the journal *Energy Policy*, Alvin Weinberg, a U.S. pioneer in nuclear safety, and colleagues called CCS "a Faustian bargain." Weinberg, who died in 2006, concluded that CCS would "make the whole of humankind more dependent on fossil fuels, and thus make a change-over later more difficult." In 2009, a group of Swedish scientists published their results from interviews with twenty-four CCS experts and confirmed Weinberg's prediction. The experts, all technology optimists, admitted that the costs of CCS remained uncertain and could escalate. Many found it difficult to justify why it was sustainable "to leave CO_2 in the ground for thousands of years for future generations to worry about." Some feared the technology would become too complex to manage. A few admitted that funding for CCS might eclipse funding for green energy such as solar and wind. The Swedish researchers concluded the political and scientific sunshine on CCS didn't accurately reflect the technology's dark uncertainties and knowledge gaps. In fact, they wrote, the urgent need for carbon-fighting technologies shouldn't be taken "as an excuse for excluding uncertainties."

Science, today obsessed with engineering the entire planet, offers big plans for the future. To battle global warming, German and Canadian scientists have proposed seeding the atmosphere with sulfur dioxide to deflect sunlight the same way volcanic eruptions do. But the project would cost billions a year and mess up the ozone layer. Others have proposed medically redesigning humans so they eat less meat and resemble Hobbits, in order to weather climate change. Interventions, says Matthew Liao, would include "the pharmacological enhancement of altruism and empathy." The project comes with one disadvantage: "We lack the necessary scientific knowledge to devise and implement geoengineering without significant risk to ourselves and to future generations."

Big thinking has also invaded the world of renewable energy. Many physicists believe that solar power can be farmed from space on a large platform and then transmitted wirelessly to global consumers. They admit that the cost might be truly nuclear but say the science is feasible. Although it would be more complicated and expensive than the invasion of Iraq, Martin Hoffert, a professor of physics at New York University, thinks the application of military-industrial complex thinking to solar energy is necessary. "We run on energy like Rome ran on slavery," he says. "But we've hit an economic, energy and environmental wall. Space-based solar power is a technologically ready path over the wall to sustainable high-tech civilization on Earth."

As scientists entertain bigger and costlier energy fantasies, the oil and gas industry maintains that technology will effortlessly extend the life of hydrocarbons. Yet the largest and most powerful industry on the planet is almost as adverse to science and innovation as nineteenth-century slaveholders were to abolition. Oil and gas companies spend only .3 percent of sales on basic research and development, less than almost every other business sector on the planet. As Philip Verleger, a smart U.S. oil economist, notes, energy companies "discover, transport, transform, and deliver energy. In general, they profit, not through ingenuity, but through commodity price increases." Corporate energy investments in science actually fell by 50 percent between 1991 and 2003. As a consequence, most of the technology currently advertised by industry as "innovative" is anything but. The inanimate slave technologies of hydraulic fracturing, in situ bitumen recovery, and deep-sea drilling are all over thirty years old. U.S. government research on energy has also declined dramatically; it now amounts to only .03 percent of GDP. A 2006 University of California study by researcher Gregory Nemet found declines in both government and private-sector investment in energy science particularly alarming "if we are to employ an innovation-based strategy to confront the major energy-related challenges society now faces."

JOSEPH TAINTER, an anthropologist who has studied why civilizations collapse, argues that modern petroleum culture has entered an "energy-complexity spiral." Although fossil fuels promised scientifically better living (comfort and ease), their energy has created so many mechanical slaves that the average North American household cannot manage the complexity, let alone the speed of these ever-accelerating servants. The cost, maintenance, and social impact of cars, iPods, televisions, furnaces, freeways, cellphones, and talking toilets have become a form of emotional and spiritual taxation, says Tainter. A typical Roman family commanded half a dozen human slaves. In contrast, Tainter now estimates that the average North American family employs four hundred inanimate slaves. Thirty slaves alone emerge from the electrical outlets that power labor-saving or time-wasting gadgets. When that many slaves structure a world in which there are "too many things to do and not enough time to do them," then a hellish level of complexity has surely arrived, says Tainter. And "a more complex society is more costly."

Society has grown weary climbing the spiral staircase, Tainter says. "Complexity grows because we have extra energy, complexity grows because we must solve problems, and complexity requires that energy production increase still more... It is a spiral that we lived with for two centuries or more. So far we have coped with this spiral fairly well. We have been able to increase production of energy and other resources sufficient to meet demand and to address the problems that we choose to solve." But as society increasingly turns to extreme hydrocarbons in the form of offshore oil, bitumen, and shale gas, energy has become more expensive and more environmentally costly. As a consequence, technologies dependent on cheap energy grow bigger and more fragile.

The alternatives we have today are limited, Tainter concludes. Society can voluntarily pump less oil and support less science. It can control the demand for energy with taxes or simply ration oil. Or it can reduce population and pray for "technological solutions." Society's elites, like most oil companies, prefer climbing even if it results in a heart attack. "To increase complexity on the basis of static or declining energy supplies," Tainter says, "would require lowering the standard of living throughout the world. In the absence of a clear crisis very few people would support this. To maintain political support for our current and future investments in complexity thus requires an increase in the effective per capita supply of energy—either by increasing the physical availability of energy, or by technical, political, or economic innovations that lower the energy cost of our standard of living. Of course, to discover such innovations requires energy, which underscores the constraints in the energy-complexity relation."

Tainter's conclusions are not discussed in petrostates or among oil lobbyists. But voluntarily consuming less does not put an end to challenges. Future problems, from soil depletion to urban crowding and food shortages, will demand more energy and complexity, not less. "We will learn this century," Tainter predicts, "whether non-fossil-fuel energies can provide sufficient energy to solve societal problems, and flexibility to increase energy rapidly when needed." Nor does he think we will stop using fossil fuels "until we are forced to."

Nearly a hundred years ago, Frederick Soddy warned that mining hydrocarbons and fueling inanimate slaves had resulted in "volcanic rather than a normal healthy growth." And this eruption came from a "stagnant pond... being drained at an ever-increasing rate and in an ever-increasing number of ways." Given the libertarian spending of energy and the repeated failure of scientists to accept "the responsibility for the profound social upheavals which their own work has brought," Soddy thought a reckoning was coming. On that day, he predicted, scientists would finally start to talk more about conservation and less about development.

10

The Petrostate

. . . . .

"Oil kindles extraordinary emotions and hopes, since oil is above all a great temptation. It is the temptation of ease, wealth, strength, fortune, power. It is a filthy, foul-smelling liquid that squirts obligingly up into the air and falls back to earth as a rustling shower of money."

RYSZARD KAPUŚCIŃSKI, *Shah of Shahs*, 1982

ONE OF THE first things that amazed sociologist Fred Cottrell about the rapid infusion of energy into North American society was how it atomized the family and strengthened the state. After forcing parents and grandparents into factories to make cheap goods alongside armies of energy slaves, governments offered schools, hospitals, and welfare as substitutes for the lost familial services. The exponential increase in mechanical serfs and their cheap consumer products also required governments to hire inspectors, scientists, and other managers. The riotous increases in capital created by this frantic energy spending required monitoring by state banks, revenue collectors, and specialized financial officers. All in all, high flows of energy not only increased the complexity of society but concentrated the power of the government to manage the volcanic eruption. Call it the Rockefeller syndrome.

But oil did even more than that. Revenues from the world's first trillion-dollar industry created a dysfunctional and novel political species that political scientists now call the petrostate. Most of the world's bizarre and highly individualist petroleum kingdoms owe their creation to either profligate U.S. petroleum consumption or the lucrative presence of black gold. Since the 1930s, every North American car owner, by buying foreign oil, has unwittingly played a role in revolutions, corruption, and autocratic government around the world. The United States, the world's first and most grandiose petrostate, brazenly set the tone. Texas, Oklahoma, Louisiana, Alaska, and California (all red states save for the Golden one) behave today like their wonky Middle Eastern brethren. Their extreme energy politics dominate U.S. culture. In fact, the petrostate—aided by an oily cult of Divine Providence—has divided the United States, as well as the world's nations, into three uneasy categories. When it comes to oil, there are only masters, traders, and slaves.

The oil industry occupies a place on the planet like no other. Every year it extracts more than $2.3 trillion worth of oil and gas from the ground, making it the world's most profitable enterprise. (U.S. oil companies alone pull in about $3 billion in profits every week.) In volume, value, and carrying capacity, petroleum and related fuels occupy a greater percentage of global trade than any other product. Black gold represents not only the largest artificial flow of energy but the single greatest transfer of wealth. Even oil men ask which will run out first—oil or money?

Oil income, whether open or secret, provides the major source of revenue for more than ninety countries. Every year, many of these governments give much of the money back, through $700 billion in subsidies to gasoline buyers and oil drillers. (The OECD estimates that the elimination of these subsidies would reduce global greenhouse emissions by 10 percent.) Seven of the world's largest corporations measured by revenue are private or state-owned oil firms. Two of the seven are owned by the Communist Party of China. Five oil firms control the majority of gasoline sales, refinery production, and exploration in the United States. In the balance of payments and exchanges between nations, petroleum reigns supreme. Regions that sell oil, such as Russia and the Middle East, record increasing surpluses, while those that buy oil, such as North America and Europe, post rising debts.

Oil exercises other destructive forms of dominion. Almost every commercial product, including toothbrushes, asparagus, and computers, comes embedded with oil. Global oil consumption grew from 35 million barrels a day in 1965 to 84 million in 2009. Americans, who consume one-fifth of the world's oil, spend about 70 percent of that allotment feeding mechanical transportation slaves, including cars, trucks, and airplanes. The drilling and refining industry creates more liquid and solid waste than any other sector on the planet. The U.S. oil industry produces more barrels of salty, radioactive wastewater than it does oil and spills, on average, 38,000 tons of hydrocarbons into the ocean every year. The industry is among the world's largest industrial consumers of fresh water. Fugitive emissions are so prevalent that more than 50,000 barrels of oil escape into water or land every day. (That's the equivalent of 450,000 escaping human slaves.) Louisiana State University's Paul Templet estimates that the oil refining business releases more pounds of pollution per job (1,048) than do papermakers (460), plastic manufacturers (222), and tobacco producers (61). The industry's drilling practices and pollution have severely damaged the Persian Gulf, the Gulf of Oman, and two of the world's greatest river deltas—the Niger and the Mississippi.

The term *petrostate* originates with political scientist Terry Lynn Karl. One of the United States' most innovative political thinkers, Karl began her career studying OPEC in the 1970s. Domestic U.S. oil production had peaked, and the fledgling industry monopoly was exercising its newfound muscle with an oil embargo that had U.S. motorists lining up at the pumps. Oil-exporting nations had created the cartel a decade earlier with the express goal of securing a fairer share of the profits for their people. Until then, only Venezuela had secured a 50-50 profit-sharing agreement with multinational oil firms.

During an interview with Karl, Venezuela's Juan Pablo Pérez Alfonso, OPEC's feisty founder, called OPEC boring and suggested Karl look instead at how oil had changed the character of nation-states. "Look around you," he told her. "Waste, corruption, consumption, our public services falling apart... The conditions of our lives are not better. We are drowning in the devil's excrement." Oil, he predicted, "will bring us ruin."

Haunted by Pérez's words, Karl began to systematically investigate the impact of oil on countries as diverse as Norway, Algeria, Venezuela, Kuwait, Nigeria, and Saudi Arabia. What she discovered infuriated the oil industry and eventually became what *Time* magazine called one of the "ten ideas that are changing our world." Everywhere Karl probed, she found that oil reduced economic diversity, fostered inequality, and sponsored autocratic government.

Karl based her seminal studies on the long history of what economists now call the "resource curse." The record shows that resource-poor countries generally perform better than commodity-rich ones, because scarcity breeds innovation and resilience. In the sixteenth century, the Netherlands, a country with limited means, economically outstripped gold-flush Spain. The great Canadian economic historian Harold Innis observed in the 1930s that one-crop economies, such as Canada's fur or forestry trades, can monopolize the interests and behavior of the state. He called it "the staple theory."

Karl's groundbreaking work took things further. Oil does not grow democracy in oil-rich, low-energy nations, she found, because it concentrates too much wealth in the hands of a few elites. As in the gold booms of old, black gold sets off an economic fever that also concentrates power. In her landmark 1997 analysis *The Paradox of Plenty,* Karl dissected the psychology of oil-exporting nations. Once oil accounts for 30 percent of a nation's exports, the money petrolizes the place, the same way human slavery changed Rome, Brazil, and the U.S. South.

Karl found that all petrostates display similar traits. Oil booms engender not only spending mania but poor statecraft, ineffectual tax regimes, political extremism, and long periods of authoritarian rule. Oil-exporting nations, she wrote in a companion essay, "The Perils of the Petro-State," "rely on an unsustainable development trajectory fueled by an exhaustible resource—and the very rents produced by this resource form an implacable barrier to change." She noted few exceptions. Even nations such as Norway exhibited signs of dangerous political petrolization. Industry apologists typically absolve oil revenue of any responsibility in the corrupting of petrostates. Former U.S. vice president and oil-patch executive Dick Cheney, for example, explains that "the good Lord didn't see fit to put oil and gas reserves where there are democratic governments." But scholars such as Michael Ross at the University of California say that's nonsense: "These countries suffer from authoritarian rule, violent conflict, and economic disarray because they produce oil—and because consumers in oil-importing states buy it from them."

Flush with dollars in the form of rent and royalties, a petrostate almost immediately lowers taxes. In so doing, it makes a mockery of one of the world's earliest cries for democracy: "No taxation without representation." In fact, nothing characterizes a petrostate more than its disdain for taxes. In oil-exporting nations from Iran to Saudi Arabia, citizens pay little or no direct income or sales tax. The United Arab Emirates has the lowest tax rates in the world. Texas, Wyoming, and Alaska, true petrostates, have no income tax. Louisiana's property taxes are among the lowest in the United States. Alaska has no sales tax. Alberta, Canada's singular petrostate, imposes no sales tax either, and its corporate and income taxes rank with the lowest on the continent. Petro-citizens possess a startling apathy and indifference to political affairs and a disconcerting loyalty to their petro-masters. This type of tax structure, noted the 2004 UNDP *Arab Human Development Report*, "minimizes the opportunity for citizens to protest against their government," because they simply haven't paid for anything. Wherever petroleum pays government bills, states serve oil's development and ultimately stop representing their citizens.

Petrostates also spend like gamblers during oil booms. This habit puts their citizens on a financial roller-coaster ride created by the world's most volatile commodity. Mexico, Nigeria, Saudi Arabia, the United Arab Emirates, Texas, Indonesia, and Alberta can all attest to the volatility of oil. Wherever schools, roads, and hospitals have been built with petrodollars, these services will shrink or fail with falling oil prices. The petrolization of public services translates into public unreliability.

Petrostates make so much money from oil—between $20 billion and $200 billion a year—that they become proverbial honeypots. That means that corruption defines life there as predictably as do leaking pipelines. "People rob," one finance minister of an oil-exporting country remarked, "because there is no reason not to." These states also devote a big share of the petro pie to their military. According to Karl, military spending in the average developing country accounts for about 12.5 percent of the annual budget. Not in petrostates. Ecuador devoted 20 percent of its budget to military spending in 2007. Saudi Arabia has spent up to an astounding 35.8 percent. The former Soviet Union, which provided cheap oil to the entire Eastern Bloc and Cuba, spent about 30 percent of its annual wealth on missiles and tanks. Notes Karl, "The extent of militarization is stunning. In the decade from 1984–1994, for example, OPEC members' share of annual military expenditures as a percentage of total central government expenditures was *three times* as much as the developed countries, and *two to ten* times that of the non-oil developing countries." Since it became a major oil exporter a decade ago, Canada has steadily increased its military budget. It now spends 18 percent more annually on its military forces than it did during the Cold War: $22.8 billion. In the United States, military spending has risen, in constant dollars, from $45 billion in 1946 to $700 billion a year today.

Another characteristic shared by petrostates is the "Dutch disease." The Netherlands experienced this ailment when it discovered natural gas offshore in the 1970s—hence the name. The first symptom is an inflated national currency that devalues and diminishes the work of local manufacturers and agricultural exporters. The more oil or hydrocarbons dominate exports, the more hollow a national economy becomes. Like a suburban family ensconced in a McMansion, a petrostate grows nothing and imports everything. In almost every petrostate Karl studied, the manufacturing and agriculture sectors withered during oil booms and rarely recovered. After the former Soviet Union used its oil to move 80 million farmers into cities, the country was forced to become a grain importer.

Dazzled by the ease of petroleum income, petrostates employ vast armies of foreign workers to do their nation's dirty work. The Saudis, for example, import nearly 10 million itinerant workers from Asia to build homes, serve meals, and clean toilets. (The treatment of maids in the kingdom is positively Roman: unruly staff have even been beheaded.) The United States, even as a declining petro force, still imports millions of Mexicans to work on its farms and in its homes. Alberta, the world's heavy-oil capital, boasts a population of only 4 million but employs more temporary foreign workers on a per capita basis than does the United States. People with oil can afford slaves.

Oil money also allows political parties, tribes, and families to concentrate power, buy loyalty, and marginalize dissidents. Petroleum buys political longevity, too: the Democratic Party ruled Texas for ninety years. The autocratic and corrupt Institutional Revolutionary Party (PRI) governed Mexico for nearly seventy years with the help of the country's formidable oil wealth. The shah of Iran promised to construct a "Great Civilization," but to most ordinary Iranians, the country became more like an Ohio suburb—one with secret police monitoring every phone call. Oil dollars cemented the long reigns of Suharto in Indonesia, Saddam Hussein in Iraq, and Hugo

Chávez in Venezuela. In Libya, oil kept Colonel Muammar Gaddafi in power for forty-two years. Alberta, the Canadian province that supplies the United States with tarry bitumen, has been a one-party state for over forty-one years. Oil can grease the wheels of political nepotism much longer than any other resource.

In sum, petrostates fund political operas and shun transparency. They uniformly abandon statecraft—the skill of making smart decisions with scarce resources on behalf of the public—to act like Las Vegas gamblers. They assume that oil wealth will paper over every policy mistake, bureaucratic folly, and systemic corruption. Finding a competent petrostate is about as miraculous as finding a polar bear in Saudi Arabia. "Countries that depend solely on oil for their livelihood eventually become among the most economically troubled, the most authoritarian, and the most conflict-ridden in the world," writes Karl.

Consider the plight of Nigeria, the world's seventh-largest oil exporter. About one in ten Americans fuel their motorized slaves with Nigeria's low-sulfur oil. But Nigeria is what University of Berkeley geographer Michael Watts calls an "archetypal oil nation." Once the oil boom began in the Niger Delta in the 1960s, the country's nascent democracy faltered. A country once fueled by agricultural exports quickly became a petrostate dependent on oil exports for 80 percent of its revenue. Watts notes that 85 percent of Nigeria's oil wealth went to 1 percent of the population, while the proportion of Nigerians living on less than a dollar a day rose from 36 percent to 70 percent. By some estimates, as much as $100 billion of the $400 billion earned by Nigeria since 1970 has disappeared into the pockets of the corrupt. Even the International Monetary Fund conceded in 2003 that "oil did not seem to add to the standard of living" there. In 1984, economist Sayre Schatz observed that Nigerian oil had created an inert economy: "For the most vigorous, capable, resourceful, well-connected and 'lucky' entrepreneurs (including politicians, civil servants, and army officers), productive economic activity... faded in appeal. Access to, and manipulation of, the government-spending process" became "the golden gateway to fortune." In Nigeria they call it "pirate capitalism."

Petrostates produce other social emissions. The rapid flood of petrodollars co-opts government agencies, which then make extraordinary deals with private or state-owned oil companies. Because oil is a capital-intensive commodity that requires its own special caste of engineers, petrostates often fall into a master-slave relationship with the resource's developers. The Total Group bankrolled Burmese dictators, while Shell funded military goons in Nigeria.

In the United Kingdom, oil wealth underwrote the success of Margaret Thatcher, the doyen of conservatism and a diligent petrolista. The Iron Lady came to power a decade after the discovery of light crude in the North Sea. The find not only increased the value of the pound but enriched a near-bankrupt British treasury by tens of billions of dollars; at one point the North Sea was producing more oil than Iraq, Kuwait, or Nigeria. Between 1979 and 1990, Thatcher used the money to lower taxes, privatize government services, attack unions, fight a war in the Falklands, and fund her right-wing U.K. "revolution." Britain's manufacturing and agricultural sectors stagnated while the banking industry soared. Thatcher, whose son was later implicated in an African oil coup plot, notably called oil "God's gift to the economy." But like many petrolistas, she saved not one penny for the future, and her political legacy for England has been bleak. A 2009 editorial in the *New Statesman* noted that Thatcher had "created, with bloodshed and war, a thin-spun, debased consumer society, the engines of which were vacuous acquisition and an obsession with celebrity. That remains the case today."

Norway (along with California) remains a global exception to the dismal petrostate syndrome, thanks to its democratic institutions and a singular accident of immigration. When the equitable nation discovered oil off its shores in the 1960s, it possessed a modest yet highly diverse fishing, shipping, and agricultural economy. Many of its traditions of competent government dated back to 1690. The world's most powerful industry faced in Norway a strong democratic ethic defended by a class of well-trained civil servants. But the scale of the jackpot was crudely surreal: the value of offshore oil promised to enrich its Norwegian owners with a lump-sum payment of $25 million, or an annual dividend of $1.25 million for each Norwegian.

With the help of an Iraqi petroleum geologist, Farouk al-Kasim, the Norwegian government crafted a unique petroleum policy. Al-Kasim, a former executive of the Iraq Petroleum Company, moved to Norway in 1969, together with his Norwegian wife, to secure medical treatment for a child with cerebral palsy. The government hired him on the spot. After a stint as chief geologist, Al-Kasim wrote a white paper on how to set up an oil industry for the people's benefit. It said go slow and save the money. He then set up and headed an independent regulator (the Norwegian Petroleum Directorate) that mandated moderate development and cleaner extraction standards. In 1991, the government created a pension fund, now worth $550 billion, for the day the country runs out of oil. The Norwegian government resisted the temptation to fuel the state with petrodollars, keeping the country going largely on domestic taxes. As a consequence, it still represents Norwegians.

Yet in some respects even Norway has fallen prey to oil's mastery. The publicly owned Norwegian oil company, Statoil, has become so rich and powerful that critic Helge Ryggvik calls it "a state within the state." Oslo has sprouted an unseemly share of McMansions built by imported Pakistani labor. Over time the oil industry has effectively subverted the Norwegian policy specifying a "moderate pace of development," draining all of Norway's largest fields when oil prices were low. The industry's constant drilling has "produced an oil-industrial complex in many ways just as dominant in the Norwegian context as the military industry had ever been in the USA," notes Ryggvik. The state's much-vaunted pension fund has also been investing in the debt-ridden oil-importing economies of North America and Europe. In spite of being one of the world's largest promoters of rain-forest preservation, Norway has failed to meet its own carbon pollution targets. Its oil-fueled salmon factories contaminate ocean waters with deadly viruses. And the world's so-called peacekeeper has also become the globe's eleventh-largest arms dealer.

After touring the Middle East and Russia in 2008, Norwegian journalist Simen Sætre categorized some of the extraordinary excesses oil had spawned. "In Turkmenistan they are building a city made of marble. The Kuwaitis work approximately 8 minutes per day. In the Gulf they are building the world's tallest skyscraper... Oil makes people believe that the desert can turn green, that socialism can be reborn, that wealth can be generated without work, and that there are no limits to where one can go." One member of Abu Dhabi's petroleum elite had his name carved on an island; the letters stretch for two miles and are visible from outer space. The same billionaire owns two hundred vehicles, including seven luxury Mercedes painted in rainbow colors. Petromania caused the Saudis to build the world's largest airport, at a cost of $22 billion ($16 billion went to pure graft). In Dubai, where temperatures average 108 degrees Fahrenheit, petromania constructed a billion-dollar indoor ski hill that uses the equivalent of 3,500 barrels of oil every day. Even the Norwegians spent $800 million on an opera house containing Italian marble. The *Apollo* moon project, a $10 billion effort, was probably the high point of petromania in the United States.

Petrostates tend to be strongly patriarchal, a trait that has nothing to do with Islam. Oil wealth and inflated petrocurrencies do away with traditional forms of female employment in local farming and small export trades while strengthening male sectors such as construction. The social consequences have been startling, says University of California political scientist Michael Ross: "It leads to higher fertility rates, less education for girls, and less female influence in the family." With fewer earned dollars in their pockets, women have less political influence. Women with fewer economic opportunities also are "more likely to support fundamentalist Islam," says Ross.

Most petrostates consolidate their political power by bribing citizens with cheap oil. Saudi Arabia, China, Malaysia, and most Middle Eastern states subsidize gasoline and diesel fuel. During Gaddifi's reign, Libya charged consumers 50 cents a gallon; Venezuela has offered gas at 12 cents a gallon. For most of the twentieth century, Americans paid less than dollar for a gallon of gas, often no more than 25 cents. At today's average price of $4 a gallon, U.S. gas prices are still among the lowest in the industrial world. (German prices are over double that.)

In 2008, a group of academics set out to discover if the devil's excrement had befouled U.S. politics. After studying both resource-rich and resource-poor states for the period from 1929 to 2002, they trained their focus on Texas and

Louisiana, two of the U.S.'s bona fide petrostates. Not surprisingly, in both states, they found politicians trying to extend their rule with petrodollars by lowering taxes and catering to the oil industry. Researchers Ellis Goldberg, Erik Wibbels, and Eric Mvukiyehe found "that oil rents appear to be politically conservative: They allow political elites to maintain control over the levers of power. Thus, oil production does appear to be undemocratic, if by that one means the opposition is less likely to come to power. We also find that resource wealth reduces economic growth."

Unlike in many petrostates, much of Texas's oil wealth has been pumped from privately owned land. As a result, billions in royalties and lease money has gone to individuals. Although private ownership tempered the resources impact on public institutions, Texas remains an oil kingdom committed to low taxes and minimal public services. Its impact on national politics has been nothing short of grand. With oil money, the state sent four presidents to the White House, including two prominent oilmen: George H. Bush and George W. Bush. The latter received more financial backing from the oil industry than any other president and appointed at least thirty industry lobbyists and executives to key government positions. To this day, Big Oil and in particular ExxonMobil largely fund the Republican Party.

Louisiana cartoonishly illustrates the power of oil wealth too. Since oil burst into the poor plantation economy in 1901, it has sustained demagoguery, corruption, and pollution. Louisiana accounts for a tenth of all U.S. oil production, and its sixteen refineries refine enough gasoline to fill 800 million cars every year. Between Baton Rouge and New Orleans, three hundred petrochemical plants pollute an area that locals now call "Cancer Alley." Yet for all the tens of billions that have poured out of Louisiana's economy, the state remains cursed with paradoxes. In a 2010 *Washington Post* article, Steven Mufson noted that Louisiana ranks "49th among the states in life expectancy, has the second-highest rate of infant mortality, comes in fourth in violent crime, ranks 46th in percentage of people older than 25 with college degrees, and ties for second in percentage of people living below the poverty line."

Louisiana has not had a clean state election since the discovery of oil. Its most infamous demagogue, Governor Huey Long, vowed initially to fight Standard Oil's control by raising severance taxes. He promised to make "every man a king" in the oil state. Instead, the money helped Long create a political empire that, as historian Michael Signer put it, "more nearly matched the power of a South American dictator than that of any other American boss." Even after Long's assassination, his cronies won elections for decades.

Oliver Houck calls oil's impact on Louisiana both transformative and unique. "We've always been a plantation state," says Houck, an environmental law professor at Tulane University. "What oil and gas did is replace the agricultural plantation culture with an oil and gas plantation culture." Oil dominates not only state budgets and politics but everyday culture. Duck hunters have to negotiate with oil companies for annual land leases. The New Orleans Jazz Festival is sponsored by Shell Oil. The Audubon Aquarium of the Americas in New Orleans contains an oil rig stocked with fish, to show that aquatic life and oil are compatible. When the U.S. Environmental Protection Agency proposed countrywide regulation of carbon emissions, Louisiana wrote four letters of protest. One even came from the state's Department of Environmental Quality. "Oil is stupendously part of the DNA," says Houck. After BP's Deepwater Horizon 2010 oil spill in the Gulf, Louisiana politicians didn't attack the company for criminal negligence; they pilloried the U.S. president.

Alaska is another classic petrostate. Ever since oil was found in the North Slope, three oil companies (BP, ExxonMobil, and ConocoPhillips) have effectively ruled its frontier. The Alaska government draws 90 percent of its general revenue from oil and gives the industry everything it wants: light taxes and even lighter regulations. Seafood, Alaska's second-largest export, now generates less money for the state than the cigarette tax.

The state's subservient political culture has offered Alaskans considerable payoffs. Half a million citizens pay no income tax. With the introduction of the Permanent Fund Dividend, an annual check based on oil investments, every Alaskan became part of the great oil compromise. In 2008, writer Chuck Thompson noticed the paradoxes: "Alaska today isn't so much a GOP stronghold as it is an oil fiefdom. Having bought the state in 1982, the oil biz to this day continues its annual payoffs to Alaskans. By 2007, the annual PFD check issued to residents was up to almost $1,600. Never mind that the prevailing local mythology remains one of self-sufficiency and rugged individuals, the importance of independence to myths is inversely proportional to the degree to which any society has surrendered its sovereignty."

U.S. Republican senator Ted Stevens, who helped build the Trans Alaska Pipeline, exercised the unimpeachable power of a Saudi sheik for forty years. Whenever the state hit the doldrums with depressed oil prices, Stevens secured massive federal subsidies to keep it afloat. In 2006, an FBI investigation caught an oil services company bribing half a dozen local politicians in a scheme later dubbed "the Corrupt Bastards Club." The conspiracy was aimed at keeping oil company taxes low. Oddly enough, the only recent politician to challenge oil's stranglehold on Alaska was Sarah Palin. The Republican populist joined with Democrats to push through a new tax deal that secured higher oil savings for Alaskans. But after Palin's abrupt departure to pursue national political aspirations, the new governor, a former ConocoPhillips lobbyist, restored the status quo, actively reducing taxes for oil firms.

Jeffrey Sachs, the Columbia University economist who has written eloquently (though not always accurately) about the resource curse, argues that Louisiana and Alaska now mirror the unhealthy predicament of the United States, "a pretty classic oil nation." The United States is the world's number-one consumer of oil, but its federal government keeps oil cheap with low taxes and multi-billion-dollar petroleum subsidies. "Big oil plays an unnatural role in our politics," Sachs told the *Washington Post*'s Steven Mufson. "Oil elects presidents, drives our foreign policy, our domestic policy, our climate change policy... It's led us to terrible energy policies and a breakdown of regulation. We look to the Niger Delta as an example of what an oil state does to its own environment, but it's precisely what we're doing to our own environment."

Religious extremism is another potent companion of the master resource. Lyman Stewart, the president of Union Oil, wasn't the only oilman to revive fundamentalism. Saudi Arabia's petrodollars have long emboldened an extreme and once obscure fundamentalist Muslim sect known as Wahhabism. These supremacists read the Koran literally, and many believe the world is flat. Outside of Saudi Arabia, few Muslims share Wahhabi orthodoxy. Yet oil has energized this "faith of hate," and its missionary clerics now train terrorists in Pakistan and Afghanistan. Every SUV driver in the United States has unwittingly subsidized the sect's revival.

In his even-handed and often startling portrait of Texas oil tycoons, Bryan Burrough documents how U.S. oil wealth gave birth to a new form of political extremism on the North American continent. Until the 1950s, American conservatism was a marginal ideology dominated by racists and anti-semites. The right wing had no real pundits and no political base. After World War II, the so-called Big Rich, four billionaire Texas families, changed that. The Cullens, Hunts, Murchisons, and Richardsons did for the United States' peculiar brand of libertarianism what oil money did for Wahhabism in Saudi Arabia: it elevated a marginal sect into a national powerhouse.

When not riding ostriches or dating Hollywood starlets, the Big Rich expounded on their radical and often hateful views. They opposed the New Deal, civil rights, and liberals. They championed moneymaking as a form of divinely inspired freedom. They backed right-wing political candidates as far afield as Maryland and Maine and played critical roles in bankrolling the presidential campaigns of Dwight Eisenhower and Lyndon Baines Johnson. Their homespun anti-semitism and outright racism appalled much of the country.

In the early 1950s, their political extremism erupted like a gusher with their calculated endorsement of Senator Joe McCarthy. The Big Rich not only funded the Communist-hunter from Wisconsin but entertained him in their homes. So much oil money funded the witch-hunting Wisconsin senator's crusade that the Texas press routinely referred to him as the state's "third Senator."

The Big Rich also planted the seeds for a right-wing media, funding what Burroughs calls a cut-rate commentary industry with an "educational" foundation named Life Line. The foundation's radio show combined fundamentalist Protestantism with political messages that attacked any individual or group who criticized the oil industry. One broadcast lamented that socialist nurses and mistresses had undermined the moral fabric of elderly rich white men.

The Big Rich of Texas weren't the only ones to sink oil money into extreme political enterprises. J.H. Pew, director of Philadelphia-based Sun Oil Company and a fundamentalist Presbyterian, was at one time the eighth wealthiest man in the United States. Pew, a modest fellow who traveled by train to work, invested in a bewildering array of right-wing institutions and causes, including the American Liberty League, the Foundation for Economic Education, the Christian Freedom Foundation, and the John Birch Society.

The latest incarnation of petroleum's religious extremism can be found in the Cornwall Alliance, a coalition of right-wing scholars, economists, and evangelicals that promotes fossil fuels as a divine freedom. The Cornwall Alliance questions mainstream science, denies climate change, describes environmentalists as a "native evil," and supports libertarian economics. One recent declaration says carbon dioxide is not a pollutant, while another claims that "modern environmentalism demands global centralization of the control of resources." A book published by the Alliance called *Resisting the Green Dragon: Dominion Not Death* portrays environmental groups as "one of the greatest threats to society and the church today." One inflammatory passage reads, "The Green Dragon must die... [There] is no excuse to become befuddled by the noxious Green odors and doctrines emanating from the foul beast." Renewable forms of energy such as wind and solar are temporary solutions for poor people, the group says, until nuclear and fossil fuel facilities can "meet the needs of large, sustained economic development." The Alliance does for petroleum what the doctrine of Divine Providence did for Southern slaveholders: puts God on the side of the dominant energy system.

In a 2010 interview, David Gushee, a professor of Christian ethics, summarized the views of many contemporary oilmen and evangelical Protestants, exposing a religious outlook almost identical to that of Wahhabism: "God is sovereign over creation and therefore humans can do no permanent damage. God entrusted the earth to human dominion and we should not be afraid of economic development or other uses of human creativity. God established government for very limited purposes such as providing for the common defense—government should not intervene much in the workings of a free market economy. The Republican Party has taken a skeptical posture toward climate and we support that posture and that party. The media is overplaying climate change worries, at the behest of scientists who cannot be trusted anyway; it may all be a conspiracy to limit our personal and business freedoms and tax us even more. The environmental movement is secular/pagan and has always been a threat to American liberties and has always been anti-business and exaggerated environmental problems. Nice worldview, huh? I disagree with just about every word of it." Gushee, though a devout evangelical Christian, sees human-instigated climate change as a threat to creation. The convenient philosophy Gushee describes is not much different from that of many nineteenth-century U.S. slaveholders. In fact, the only defense for slavery that gained any traction at the time originated with evangelical Protestant clergymen, who wrote most of the pro-slavery tracts. "No human institution, in my opinion, is more consistent with the will of God, than domestic slavery," wrote one slaveholder. They called it the doctrine of Divine Providence.

The recent activities of Charles and David Koch, two of the United States' wealthiest men, illustrate petroleum's enduring grip on the world's first petrostate. (Their father, Fred Koch, a chemical engineer, helped Stalin build fifteen modern refineries and later regretted the deed.) The libertarian

Koch brothers own refineries and pipelines in Alaska, Texas, and Minnesota. Their companies rank among the country's top ten polluters. Charles and David have funded the Cato Institute and the Mercatus Center, both of which routinely denounce climate change and call for less government. But the Kochs' crowning achievement has been the polarizing work of the foundation Americans for Prosperity. The organization seeded, funded, and trained members of the Tea Party, which seeks to recreate the simple nation that supposedly existed in the early days of cheap oil. It's a petroleum-funded political movement designed to support more oil and less government.

In many respects, petrostates only mirror the greed and arrogance of oil companies. No one has ever accused a multinational oil corporation of being ethical or transparent. Oil companies routinely defraud states and citizens alike. Big Oil not only lies about pollution and other matters but routinely attacks critics and competitors with smear campaigns and lawsuits. (America's litigious reputation probably started with Big Oil.) Every major U.S. oil regulatory agency, from the Texas Railroad Commission to Minerals Management Services, has been compromised by scandal. Like petrostates, they are captured and conflicted entities. Oil firms, which typically make more money for less effort than any other industrial sector, are also as volatile and spendthrift as petrostates. Subject to constant bust and boom cycles, the industry is forever firing employees or hiring new batches of contractors. (The employees of oil companies make up an energy aristocracy and are among the best paid workers in the world.) Whenever citizens object to chaotic drilling, hydraulic fracturing, water contamination, or environmentally devastating pipeline routes, the oil companies simply pour money into local community organizations, build sports facilities, and fund Republican candidates in the United States. Like petro autocrats, they crudely purchase consent and studiously buy silence. Not surprisingly, the industry scores the lowest ratings of any business sector in Gallup's ten-year history of measuring public opinion on U.S. corporations.

MORE THAN THIRTY countries around the world now derive at least 30 percent of their income from oil and gas production. Their well-armed elites are richer than eighteenth-century slave-plantation owners, and their poor eat less than Roman slaves. Even in the few nations that discovered oil as secure democracies, oil wealth has eroded institutions and principles. The petroleum *dominus* can have but one master. Just as adding a spoonful of sugar to her tea connected an eighteenth-century London matron to the bloody and deadly slave trade, so the purchase of gasoline today ties every car driver to petro-kingdoms, poisoned waterways, and political corruption. The terrible logic of petroleum boils down to one truth, says Terry Lynn Karl: "It is simply a lot easier and faster to build a pipeline than an efficient and representative state."

Two prominent Russian physicists, Anastassia Makarieva and Victor Gorshkov, recently analyzed the world's new petroleum energy order. "The developed world has literally become the economic slave of the domestic as well as, for the most part, of foreign energy sellers," they concluded. The physicists calculated the world's 2005 gross domestic product at around $45 trillion USD. Five trillion of that, or more than 10 percent of global GDP, went directly to energy expenditures and the petrostates. Between 2005 and 2009, an additional $850 billion went to oil-exporting countries. Ten percent of global GDP now flows to an economic sector that provides highly specialized jobs for less than 0.1 percent of the world's population.

The Russian scientists call the wealthy inhabitants of these petrostates a "vacant population." Like the absentee owners of Caribbean slave plantations, they don't have much to do. The Middle Eastern and North African oil-exporting nations have the highest economic dependency ratio in the world; each worker supports two nonworkers.

With a few exceptions (Norway again), oil exporters can't be accused of spending their petrodollars on making the world a wiser place, either. In 2002, for example, Saudi Arabia and Kuwait spent less on science research than most of the world's poorest countries. In contrast, an average of 7 percent of both countries' GDP went to guns, bombs, and aircraft. "Generally," concluded the Russian physicists, "the activities of a vacant population are economically uncompetitive; at the worst—directly aimed at undermining the base of civilization (terrorism)." (Curiously, due largely to lack of opportunity and the profession's conservative bent, more engineers become Islamist extremists than any other professional group in the Middle East.) Makarieva and Gorshkov argue that this prolonged "vacant" condition in oil-exporting nations amounts almost to "exclusion from active participation in the scientific and technological progress" of civilization. Moreover, "it results in decline of educational standards, disappearance of qualified specialists and general social degradation."

The physicists, who freely acknowledge their membership in Russia's dysfunctional petrostate, go even further in their analysis. Like most critics of modern economic thinking, they recognize that "dollars and Joules are different units of one and the same dimension, energy." But given that oil importers, on average, pay forty times the real cost of extracting petroleum, the resource's slaveholders promote grave global fiscal imbalances: "The money the owner receives does not correspond to any work performed, because the cost price of this production is negligibly small compared to the market price. In other words, there appears an infinite flux of money (which is equivalent to a flux of mass or energy) out of nothing." As developed nations give away more and more of their hard-earned economic surplus to the petro-masters, their ability to address pressing environmental and climate problems measurably declines. "Under the burden of dramatically overpaying for energy, the economies of developed countries already function on the verge of losing their integrity."

The Russians conclude by offering a novel solution. To end rising energy prices and the unbalanced flow of capital to petrostates, they propose controlled global pricing for oil at the cost of production. Such a policy would smash the monopoly of petrostates and oil companies. It would also free up one-tenth of the world's GDP to be spent on other goals, such as preventing the deforestation of tropical countries. But it's unlikely that the inertia of the present system or the world's most powerful companies would allow such dramatic reforms. Instead, the Russians predict more slave revolts. "Psychologically, citizens of the developed countries will find it extremely difficult to admit their current status of economic slaves," say Makarieva and Gorshkov. Once they do, however, the physicists foresee the day when these well-armed slaves will attempt to free themselves from petroleum slaveholders "who do not yet have nuclear weapons." It's a frightening scenario.

Petroleum's shadow is not confined to the Middle East. Canada has abandoned climate-change commitments, denounced environmentalists as "radicals," and cut both scientific and environmental research. But perhaps no political relationship better illustrates petroleum's shifting master-slave dynamic than that of the United States and Saudi Arabia. The United States brought the oil trade to the desert in the 1930s. The kingdom, which did not officially abolish human slavery until 1962, learned well from the American pioneer. When the United States fell from grace in the 1970s to become just another oil-importing nation, Saudi Arabia's wealthy sheiks assumed the role of master. Since then, successive U.S. presidents have arrived in Riyadh with caps in hand. When Ronald Reagan needed money to fund pro-American terrorists in Nicaragua, he went to a sheik. Bill Clinton once asked a sheik for $30 million for an American university. George W. Bush traveled to Riyadh in 2007 and begged like a slave for lower oil prices. "My point to His Majesty is going to be, when consumers have less purchasing power because of high prices of gasoline," he said in an interview. "In other words, when it affects their families, it could cause this economy to slow down. If the economy slows down, there will be less barrels of oil purchased."

To a proud desert people, such slavish behavior appeared unseemly. King Fahd bin Abdul Aziz allegedly described U.S.-Saudi relations this way in 1993: "I summon my blue-eyed slaves anytime it pleases me. I command the Americans to send me their bravest soldiers to die for me. Anytime I clap my hands a stupid genie called the American ambassador appears to do my bidding. When the Americans die in my service their bodies are frozen in metal boxes by the U.S. Embassy and American airplanes carry them away, as if they never existed. Truly, America is my favorite slave."

The cost of this overt slavery has nearly bankrupted the United States. The geographer Roger Stern calculates that between 1976 and 2007 the U.S. military spent an average of $225 billion a year solely on oil-supply protection in the Persian Gulf, more than it did during the Cold War with the Soviet Union. Deploying aircraft carriers in the region cost the United States $7 trillion. "This substantial military investment is not a remedy for the market failure at the heart of the regional security problem, which is oil market power," wrote Stern. Instead of this show of force, the United States should have pursued an aggressive course of political emancipation by reducing energy demand at home. But given the power of the oil companies, such liberating strategies have yet to be adopted. It was clear to geologist Earl Cook, writing nearly thirty years ago in *Man, Energy, Society,* that the master producers of oil will ultimately dominate. "The greatest danger in our bemused drift towards the energy waterfall," Cook wrote, "is that the resulting shock will find us stripped of democratic government by an opportunistic group that comes out on top in the wreckage, a group that controls us through their control of the energy systems."

11

The Surplus Devolution

. . . . .

"The line between failure and success is so fine that we scarcely know when we pass it: so fine that we often are on the line and do not know it."

The Electrical Review, 1895

IN A 1795 series of letters on art and beauty, Friedrich Schiller, the German historian and poet, defined play as "the aimless expenditure of exuberant energy." He thought that artists, children, and animals played when there was more food on the table. "When hunger no longer torments the lion, and no beast of prey appears for him to fight," he wrote, "then his unemployed powers find another outlet. He fills the wilderness with his wild roars and his exuberant strength spends itself in aimless activity." British philosopher Herbert Spencer later borrowed the notion and called it the "surplus energy theory."

A surplus is what hunters and gatherers captured from fish-bearing rivers and nut-laden trees; it's what early civilization expropriated from the sweat of slaves; and it's what modern society fritters away in fueling billions of mechanical servants. Ralph Waldo Emerson, another poet, often said in his lectures that what a nation did with its "surplus produce," or energy, defined its character: "One bought games and amphitheatres; one, crusades; one churches; one, villas; one horses; one, operas; one, tulips; ours buys railroads, ships, mills, and observatories." Emerson, who lived at a time when muscle, wind, and animals still supplied most of the United States' energy, would have been appalled both by how his country eventually spent its oil and by how quickly it exhausted its surpluses.

Energy experts largely avoided the topic of surplus in the first half of the twentieth century. But sociologist Fred Cottrell, writing in the 1950s, attributed America's exceptional development to the nation's bountiful energy profits. Societies that picked high-energy berries expanded, explained Cottrell, while those that chased low-energy rabbits declined. "A stroller eating blackberries growing wild along the road" got back more energy than he spent. "On the other hand, a man who runs down a jackrabbit in an 80-acre field" would burn more calories than he captured. Low-energy societies that lived on sunlight, Cottrell argued, paid close attention to their gains and losses because energy debt invariably led to cultural disintegration. But in societies hooked on petroleum, wrote Cottrell, "the facts are extremely difficult to come by." In fact, oil produced such rich surpluses that most people stopped thinking about energy altogether.

No recent academic has analyzed the significance of energy surpluses and net gains better than Charles Hall. In the 1980s, the plainspoken New England systems ecologist came up with the formula for energy return on energy investment (EROI). It's a biological equation devoted to inputs and outputs, and it goes like this: "EROI = energy gained divided by energy required to get that energy." The formula can also be expressed as "Energy returned to society divided by energy required to get, deliver, and use that energy." If a civilization, a tribe, or a corporation secures more energy returns from an activity than it invests, then it enjoys high EROIs and has time to play. But when a nation—or, say, a salmon—uses more energy than it gets back, trouble or starvation follows. Another way to convey Hall's measurement is to consider EROI as a percentage of fuel delivered to the global gas station. A unit of energy with an EROI of 100 puts 99% of that fuel to work at society's service. A gallon of fuel with an EROI of 2 delivers only a 50 percent gain. A fuel with an EROI of 1:1 is obviously useless. That's when the car stalls on the freeway. EROIs lower than 10 march a civilization progressively towards what is called the "net energy cliff."

Why a salmon? Fish taught Hall much of what he knows about energy surpluses. The energy of fish tells a lean story. A rainbow trout, for instance, will position itself in a fast-moving stream to collect floating insects. The gamble works if the fish ingests more calories than it spends on fighting the current. The fastest-growing trout finds the ideal current, the one that delivers the most food with the least effort. A slow-growing and unlucky cousin might drift aimlessly and then die. In the world of both fish and humans, there is no life without some form of energy profit.

When Hall studied the movement of twenty-seven species of fish in North Carolina's New Hope Creek in the 1970s, he made some energetic findings. He found that larger adult species moved upstream to spawn in the more productive shallow pools. Since the stream's upper reaches also offered the adults' offspring a rich nursery, the young didn't have to spend many calories in search of food. Hall also discovered that for every unit of energy the fish spent migrating upstream, they returned at least four units to the next generation of fish. The flow of fish upstream also acted like a natural energy pipeline, delivering essential nutrients to forest and creek alike.

The ecologist next studied the tiny Pacific salmon smolts that migrated all the way to Alaska and the Aleutians. Hall wondered why these young fish didn't just park themselves at the nutrient-rich mouth of the Fraser River and enjoy a free meal. Energy profits again figured into the answer. The salmon, he discovered, followed higher densities of zooplankton that moved up the coast toward Alaska. These concentrated balls of edible fish food ensured higher energy returns, faster growth, and better survival rates. Such findings convinced Hall that the world revolves around surplus energy: "Everything in life is about energy costs and energy gains," he says.

The world's global fisheries and the petroleum industry illustrate the importance of energy profits in startlingly similar ways. These two primitive-energy industries, obsessed with capture, could almost be twins: both are testosterone driven, highly subsidized, and prone to wild busts and booms. Half of all oil is traded internationally in tankers; 50 percent of all fish landings go to distant markets. Both industries lie about the quality of their resources. Just as China has overstated its catches to promote the illusion of endless growth for twenty years and thereby obscured the decline of global fisheries, so Saudi Arabia and Shell have lied about their oil reserves, masking declining reserves of light crude. Regulators for both industries, among the most corrupt in the world, have no or little regard for conservation.

There are key differences. One industry mines the revenue of sunshine, while the other exhausts ancient capital. Fishing once netted big fish, a renewable biomass that ate small fish that dined on tiny creatures energized by photosynthesis.

Petroleum mining attacks nonrenewable stocks of stored solar energy. But tellingly, both industries began by netting the easy catches at the top of their respective pyramids. They hunted the big fish first because those provided the richest returns. While the oil industry targeted some 600 giant oil fields on land, fishers captured the largest species swimming in lakes and rivers. After cleaning out the richest pools, both industries packed their bags and moved offshore. Fisheries started the offshore trek 400 years ago; the oil business followed in the 1950s. Once the fishing industry had nearly exterminated the big fish such as ray, slipmouth, cod, and wolffish, it left the coastline for murkier, deeper waters and smaller prey. Industrial trawlers can fish to depths of 1,600 feet and lift boulders the size of Volkswagens off the ocean floor. The oil industry perforated dense shale rock with high-pressured blasts of water, sand, and chemicals to release small amounts of oil and gas or drilled miles under the ocean floor.

The global fishery catch, once thought inexhaustible, quietly topped out at 94 million tons in the 1980s and has since dropped precipitously. The oil industry, which deemed cheap oil endless, peaked at between 85 and 88 million barrels of oil a day around 2008. It can no longer grow without massive investments in costly unconventional and dirty fuels. Scientists generally agree that the world's cheap, light, and easy oil will be exhausted by 2030 just as Canada's northern cod fishery was in the 1990s.

Fisheries biologists estimate that the global fish industry has now netted and processed two-thirds of the ocean's biomass. The first steam-powered trawler appeared in 1850. But it was oil and its attendant horsepower slaves that radically magnified the power of industrial fishing. The master resource caused the number of inanimate slaves on every fishing boat to explode after WWII, explains Daniel Pauly. The Paris-born marine biologist is the world's foremost expert on global fisheries. What was once "a little boat bobbing in the water" now comes equipped with fish finders, sonar, scanners, electric winches and satellite feeds. "[These boats are] using technology designed to hunt submarines," explains Pauly. "We are using war technology against fish and the outcome is preordained. A modern fishing boat is a tank."

Yet studies done by J. Cutler Cleveland and colleagues on the New Bedford fleet, then harvesting the globe's sixth-largest catch in tonnage, showed that as fishing vessels employed more inanimate slaves to get bigger and faster, their nets delivered smaller yields. Between 1963 and 1988, while horsepower per vessel increased from 252 to 624 horsepower (allowing longer fishing trips into deeper waters), the New Bedford fleet spent more diesel to capture less edible protein. Despite burning three hundred times more fuel in 1988, the fleet actually caught 30 percent fewer fish than it had twenty years earlier. Its edible protein EROI declined fivefold, though government subsidies masked the real costs of the decline.

Charles Hall began to wonder if the oil industry too was pulling in smaller catches by spending more oil. Earlier studies by geologist Marion King Hubbert had found persistent declines. In the 1930s, the industry recovered about 250 barrels per foot of exploratory drilling, but that surplus dropped to 40 barrels per foot in the 1950s, rising intermittently when new fields were drilled. After the industry drained large oil fields in the North Sea and Texas, then moved offshore and to Arctic frontiers like Alaska, Hall suspected there would be further declines. In 1980, he assigned the task of finding out to a graduate student, Cutler Cleveland, who produced an N-shaped graph that showed energy surpluses for oil going up and down like a yo-yo. "The yield per foot of drilling would reach a minimum and then jump back up, then down even more sharply," recalls Hall. Puzzled, Hall then asked Cleveland to add the number of feet drilled per year to the study. Once he did, the graph looked like a snowball going downhill: the number of barrels recovered per foot had dropped from 50 in 1946 to 15 by 1978. "It looked just like the falling catch graph for the fisheries," says Hall. The *Wall Street Journal* reported on the findings in 1981 under the headline "Increased Drilling for Oil May Consume More Energy Than It Gleans."

Since then, Hall and his colleagues, including Cleveland and David Murphy, have refined their work on EROI. Hall considers the measurement a critical tool for shining light on the quality of energy available for a society. He notes that the global EROI for oil production delivered returns of 30 a century ago but stands today at less than 20. In 2006, the U.S. Department of Energy estimated that the U.S. oil production EROI had dropped to 10: the high energy costs of offshore drilling, heavy-oil pumping, and hydraulic fracturing ate up more surpluses. But the concept of emptier energy nets and their societal implications still gets little attention in government and academic circles. "We never got any money to do this," says Hall. "It all happened on weekends or pro bono. No government agency is interested in the information. Most science, to be honest, promises some form of candy. EROI doesn't do that and we don't do that."

In a 2011 collection of studies for the journal *Sustainability*, Hall offered a fuller picture of petroleum's shrinking energy profits. Industry trends now look as dismal as disappearing stocks of wild salmon and tuna. Hall estimates that it took only one barrel of oil to find an incredible 1,200 more barrels in 1919. In today's capital- and energy-intensive petroleum fishery, reduced to horizontal drilling and bitumen mining, industry spends one exploratory barrel to catch just 5 more petroleum fish. The energy return on the cost of producing the oil has also dropped. One barrel put another 24 in a pipeline in 1954; today the yield is 11. Oil's initial bounty and formidable EROI, says Hall, "had a great deal to do with a tremendous increase in wealth in the first part of the 20th century." But society is now living on old oil fields, and "we're spending more energy to find less and less energy," he explains. "Politicians who say, 'Drill, baby, drill' have their heads up their asses. You don't get more oil by drilling more. You just get less efficient returns. You only get more oil by drilling thoughtfully."

The world's fisheries mirror these energy declines, although cheap oil and subsidies have hidden the losses as industry compensates by moving from low-value catches like cod, once a staple for Caribbean slaves, to luxury items such as shrimp. Industry trawlers burn enough oil to support a ninety-mile line of Corvettes, bumper to bumper, with all engines revving. In some fisheries the edible protein hauled up amounts to less than 10 percent of the fuel energy burned. In addition, one-third of the world fishery catch—sardines, anchovies, and mackerel—is wasted as animal feed for industrial livestock or fish farms. Because it takes four to five pounds of small fish to make one pound of factory salmon, the EROI for factory fish hits 2 or lower. Daniel Pauly describes the fueling of fish farms with tons of small fish, which leaves less food for marine mammals as well as for people, as "the aquatic equivalent of robbing Peter to save Paul." The oceans once made salmon for free. Now it takes five pounds of oil to fatten a single pound of caged salmon in a factory operating on the same principles as a Ford assembly line.

Industrial fishing has become such a backward, corrupt, and duplicitous enterprise that reform will not come from within, says Pauly. But the end of cheap oil may offer an opportunity for change. When the price of oil hit $150 in

2008, the over-energized trawler fleet contracted. High-cost oil may restore small-scale fisheries that employ fewer inanimate slaves and respect local waters. "In some countries sail boats are now being used again for fishing," says Pauly. These reforms, combined with a substantial reduction in catches, the cancellation of $30 billion in subsidies, and the creation of extensive marine parks, argues Pauly, could prevent the collapse of the world's fisheries. Pauly, whose ancestors were slaves imported to the United States, thinks it may be too late. However, he says, "you do the right thing whether you are successful or not. That's how you give meaning to life. People in the depths of slavery fought against it because they had to. They could not do otherwise and they did not wait for emancipation. The fight against the machine is the challenge of our generation just as the fight against fascism was fifty years ago."

DIMINISHED CATCHES of oil and fish ultimately mean diminished living. Cheap oil built giant cities, factory farms, extensive roads, big science, and even bigger government on returns greater than 20 EROI. Charles Hall estimates today that business as usual probably requires an EROI of at least 5, and even that quantity of surplus might not be sufficient to keep "all the infrastructure [needed] to train engineers, physicians and laborers." He reckons that a sustainable EROI might be as high as 10. Any dirty fuel or renewable energy that provides smaller gains would translate into smaller government, markets, cities, and farms. With less surplus to energize a civilization dependent on billions of energy slaves, Hall suspects we might have to retire lots of mechanical help as well as reduce both labor productivity and wages. We might even have to reconsider how society distributes wealth.

An energy return of 10 is a significant number. Most hunter-gatherers lived on surpluses greater than that. But contrary to modern assumptions, many preindustrial societies enjoyed both high EROIs and stable living conditions. The !Kung bushmen of the Kalahari, for instance, fashioned lives that were neither nasty nor brutish. Just one !Kung adult could pick and hunt enough food to support four or five children and elders in a couple of hours. Securing food energy for the group rarely took more than two to three days of the week. The resourceful !Kung also enjoyed a much more varied diet than do the citizens of most industrial societies, including twenty-nine different fruits and berries, thirty different roots, and the drought-resistant mongongo nut. Gathering plants yielded higher returns than hunting animals and supplied 60 to 80 percent of the diet. During a sustained drought in the 1960s, the !Kung remained as resilient as ever while their agricultural neighbors starved in refugee camps. By spending a third of their time shopping in the bush, the !Kung enjoyed EROI surpluses of 12 to 1. With that heady profit they cared for the young, elderly, senile, blind, and sick and spent nearly half of the time playing, visiting relatives, and dancing. Steady leisure invariably followed steady work. Their energy returns were so profitable that U.S. anthropologist Marshall Sahlins calls hunter-gatherers "the original affluent society."

Charles Hall doesn't think many people will be able to keep bankers' hours or dance like the !Kung with declining surpluses. On an island like Puerto Rico, where the ecologist has studied energy returns in the rain forest for decades, the importance of EROI rolls sharply into focus. Before the advent of oil, the island was a poor and hardworking village that sacrificed its forests for sugar exports. But after the introduction of oil and Operation Bootstrap, Puerto Rico became an industrial marvel. Thirty percent of its forests grew back as Puerto Ricans employed their new energy slaves to make pharmaceuticals and other goods. Now that the price of oil has reached

$100 a barrel, and given that most of the island's power comes from oil-fired generators, Hall fears the clock might run backwards.

Hall thinks that EROI also explains the damnable predicament of the United States. The world's first petrostate, which once dazzled the planet with drilling energy returns of 1,000 to 1, now sputters along on returns of 10 to 1 or less. "I have seen the future, and it's here," says Hall. "Everywhere you turn in the United States you see economic constriction. About 46 of 50 states are broke. Our universities are broke. All the Tea Partiers are bent out of shape by the national debt. The country just can't do a lot of the things that it used to do. We have had little or no increase in GDP and no increase in energy use for six years. And that's not a coincidence."

EROI also provides a reality check on renewable energy. Unlike fossil fuels, which nature stored in high-density pools and formations, the free energy flows of wind and solar must be collected and concentrated by people. That effort requires a lot of energy. Biofuels—fuel oils made from crops—illustrate the conundrum. Many scientists believe that the world might someday run on factory-made algae instead of petroleum. They argue that the microscopic plant grows fast and can potentially generate great volumes of lipids, or energy products. But to date, the production process yields EROI returns of significantly less than 1 and costs about $35 to make a gallon. The heavily subsidized ethanol industry also produces meager surpluses. Using oil-driven tractors, plows, cultivators, planters, combines, and irrigators to harvest corn for "gasohol" yields a slightly positive EROI. But when climate change, soil erosion, and the rising price of food are added to the equation, ethanol goes over the net energy cliff. (*The Economist* estimates that a large tankful of ethanol represents enough corn to feed one person for a year.)

In a critical review of surpluses generated by renewables more popular than biofuels, Hall found the same troubling story. None returned the same surpluses or gains as conventional fossil fuels. Industrial wind farms generate EROIs of approximately 18 to 1, nearly three times less than coal. (Bigger turbines appear to deliver bigger gains because they can capture more wind more efficiently.) But the wind blows less than one-third of the time on average. Tidal or wave energy may produce some surpluses, but the existing projects are too few to assess their true contribution. Tapping into the earth's ambient heat via warm subterranean waters can yield returns of 5 to 1. Hydro power yields high returns, but most sources have already been dammed. The world's most expensive energy, nuclear power, seems to have paltry returns of 5 to 8, though we lack good data. Capturing sunlight, often viewed as the utopian solution, comes with challenges too. Solar photovoltaics, which require energy-intensive manufacturing, must extend over large areas to net only a fraction of the sun's energy. To date, photovoltaics offer energy returns ranging from 3 to 10.

Unlike Hall, Italian physicist Ugo Bardi believes that solar returns will increase with new and better technologies. Bardi argues that polycrystalline silicon offers an EROI of 15, while thin film cells might produce surpluses of 40. He also supports high-altitude kites, a novel technology that generates energy with yo-yo-like jerks to a generator on the ground. Scientists estimate that the kites might yield returns of 100. But tapping high-altitude winds could also affect rainfall patterns. Every energy surplus comes with some slavish and hidden cost.

Extreme hydrocarbons such as tar sands and shale oil perform worse than many renewables. They are the equivalent of the sardines, jellyfish, and slime at the bottom of the ocean food chain. Even the oil patch calls them "difficult" and "ugly" resources. The shale gas revolution makes a good but nuanced example. Now that the easy natural gas is gone, industry must drill deeper and then crack open giant shale formations as dense as concrete. The volume of methane released is so small that to make up the difference, companies must amass land bases the size of West Virginia. To blast open fractures in the rock, the industry pumps highly pressured volumes of sand, toxic chemicals, and millions of gallons of water a mile underground. In 2003, the industry used 2 million horsepower in its hydraulic fracturing operations. By 2011, it will need 11 million horsepower, or 8 gigawatts of energy. That's the same amount of power generated by eight large U.S. nuclear power plants. The EROI for shale gas starts relatively high, at 80. But that number represents a 28-point drop compared to conventional gas a decade ago, and within another decade rapid depletion rates and rising energy costs will cause shale gas surpluses to fall catastrophically.

The energy returns for Canada's bitumen—a tarry, overcooked crude mixed in sand and clay—are more abysmal. Once dug out of the ground, bitumen must be extensively upgraded with hydrogen and then put through a highly polluting refining process. The amount of energy needed to dig bitumen out of the ground, remove the sand and clay, and then convert the junk into synthetic crude ultimately delivers an EROI ranging from 3 to 5. That makes it a lowly competitor with biofuels. Deep deposits of bitumen require even more energy to access. Massive steam plants boil water with natural gas and then inject highly pressurized volumes of steam to melt the bitumen (about four barrels of steam for every barrel of bitumen produced). The inefficient process, which operates like a bottom trawler, typically heats up more rock than it does bitumen, delivering returns ranging from 3 to 1. Some bitumen steam projects actually deliver negative returns and perform as poorly as ethanol. Oil shale (kerogen), another poor and undercooked petroleum product, offers paltry net gains of 2 or 3. Its recovery process heats the ground up to 1,100 degrees Fahrenheit, with electrodes no less. Daniel Pauly compares mining unconventional hydrocarbons such as shale gas and bitumen to drag-netting *bêches-de-mer*: "Do you know what sea cucumbers eat? They lick the bottom of the ocean and they eat shit." Canada opened a dragnet sea-cucumber fishery in Newfoundland after cod stocks collapsed due to overfishing in 2003. Thirty percent of the sea-cucumber catch is water and debris. Each year its EROI declines further.

The Global Energy Systems group at Sweden's Uppsala University has added a new fish to the surplus debate. The group's researchers note that the global oil boom started from nothing in the 1870s and then grew at 7 percent annually, doubling its output every ten years. Oil offered so much concentrated surplus that it served as a "dream given form" and "one of the most extreme events in human history." Oil output even grew faster than nuclear power or hydroelectricity. By 1970, oil was delivering two thousand times more energy gains than it had in 1870. Given this unusual record, the Swedes don't think any of the renewables, with their typically lower EROIs, could ever scale up like an oil boom. "The development and expansion of alternative energy sources has started too late to produce even a significant contribution over the next couple of decades," the researchers write. They also pose a provocative question: "Should energy planning be based on the belief that new energy sources can grow faster than anything ever seen before in history?"

The implications of this analysis and Hall's EROI work are disquieting. Neither unconventional fossil fuels (dirty oils) nor green renewables offer enough surpluses to feed the world's hungry energy slaves, let alone their masters. Moreover, most scientific and government agencies no longer collect the detailed energy information necessary to make these calculations. Often they ignore the issue altogether. Charles Hall, who accuses energy regulators of criminal negligence, argues with Ajay Gupta that "there needs to be a concerted effort to make energy information more transparent to the people so we can better understand what we are doing and where we are going." The situation, he writes, grows more extreme by the day. "The EROI of the fuels we depend on most are in decline; whereas the EROI for those fuels we hope to replace them with are lower than we have enjoyed in the past. This leads one to believe that the current rates of energy consumption per capita we are experiencing are in no way sustainable in the long run. At best, the renewable energies we look toward may only cushion this decline."

Not long ago, Hall stood in the middle of Puerto Rico's Luquillo Experimental Forest to talk about the importance of energy costs and gains for the Discovery Channel. What works in a rain forest, he told viewers, also works for civilized society and for empires based on oil. "I think it's very simple," he said. "Just as the forest cannot use more energy than is available by photosynthesis, human civilization cannot use more energy than what's available from the sun or from our temporary joy ride on fossil fuels."

12

Oil and Happiness

. . . . .

"Everything in modern life is congested—our politics, our trade, our professions and cities have one thing in common: they are all congested. There is no elbow-room anywhere... There can be but one path of escape, and that is backwards."

ARTHUR PENTY, *Guilds and the Social Crisis*, 1919

EVERY OIL COMPANY and petrostate today whistles a patriarchal tune. The American Petroleum Institute says the world needs more energy because oil drives "the American dream" and gives people the freedom to move anywhere, anytime. For Rex Tillerson, chairman and CEO of Exxon Mobil Corporation, the recipe for global prosperity is simple: "We must produce more energy from all available and commercially viable resources." Pipeline builders echo that the world is "clamouring for more energy." With religious fervor, Shell executives swear that they will "produce more energy for a world with more people" so that millions can climb up "the energy ladder."

These self-serving arguments from the world's petroleum brokers are based on a singular falsehood: that more energy translates into better living. Decades of human slavery peddled the same lies. Eighteenth-century Liverpool and Bristol slave traders contended that trafficking in human energy was "the best traffic the kingdom hath"; the world needed more slaves to end global drudgery and provide the necessities of life. One 1749 pro-slavery pamphlet declared that "the most approved Judges of the Commercial Interest of these Kingdoms" had deemed slavery "most beneficial" because it employed ships and seamen. Slaves, the pamphlet said, were "the daily bread of the most considerable of our British Manufactures." A popular American defense of slavery argued that "the products of labor feed and clothe the world, and thus conduce to the welfare and happiness of mankind. Coerced labor is better than no labor." Every dominant energy system, from human slavery to nuclear power, has regarded itself as the master resource and has defended its reign with combustible rhetoric and the call for more.

Yet none of these arguments are rational, moral, or equitable. And despite the claims that high energy spending will give us all better lives, happiness research yields some startling insights into the nature of energy consumption. North Americans now use up to fifty barrels a person a year equivalent in oil, petroleum, atoms, and electricity. (Direct oil spending amounts to twenty-three barrels per person.) Americans, says University of Manitoba energy expert Vaclav Smil, "have been living beyond their means, wasting energy in their houses and cars and amassing energy-intensive throwaway products on credit." Smil believes that good health and political cheer, if not happiness itself, can be achieved on much less.

Smil, who calls himself an incorrigible interdisciplinarian, has written more than thirty books and four hundred scientific articles on energy, population, and natural resources. Even the billionaire Bill Gates finds his work illuminating. "Rising energy and material consumption," says Smil, "is not a viable option on a planet that has a naturally limited capacity to absorb the environmental byproduct of this ratcheting process." Like the eighteenth-century abolitionists, Smil considers unfettered demand and consumption a deeply moral issue.

Smil refers often in his writings to a U.S. insurance statistician and demographer by the name of Alfred Lotka. Lotka was the first to observe that all living species tend to maximize available energy for more successful living. A coral reef, for instance, does a much better job at converting solar energy into diverse forms of life than a desert does. A hardwood forest will grow ever more leafy and dense as it ages to secure more available sunlight. "In the struggle for existence, the advantage must go to those organisms whose energy-capturing devices are most efficient in directing available energies into channels favorable to the preservation of the species," wrote Lotka.

Ten thousand years ago, a hunter-gatherer collected the energy equivalent of 1.5 barrels of oil a year from plants and animals. Chinese peasants upped the ante in 100 BC with wood and coal to secure a fortune of 3 barrels of oil equivalent per capita a year. That harvest didn't change much until the Industrial Revolution. But by 1880, coal and steam slaves had exploded the amount of energy available to the average person to the equivalent of 15 barrels of oil. A hundred years later, Europeans gorged on 26 barrels per person annually. Americans wanted more, and they got it. With 40 percent of their energy coming from oil and another 25 percent from natural gas and coal, they burn through 50 more barrels per capita annually.

Given that the United States consumes twice as much energy as the richest European nations, Smil poses an impolite question: What does it get back in return? "Are Americans twice as rich as the French? Are they twice as educated as the Germans? Do they live twice as long as the Swedes? Are they twice as happy as the Danes or twice as safe as the Dutch?" The answer is a profound no. On quality of life indicators such as child mortality and educational achievement, the United States doesn't even rank in the top ten. Americans have much higher rates of obesity, suicide, murder, and incarceration than do Europeans and the Japanese. Moreover, American literacy and numeracy rates are in steep decline. And research continues to show that Americans are less happy today than they were fifty years ago. The majority of U.S. citizens now say they detest the accelerated nature of their labor alongside inanimate slaves. All of the research, in fact, shows that happiness consists in the very things traditionally denied to slaves: healthy children, close friends, a loving spouse, good health, and rewarding work.

According to Vaclav Smil, 10 million U.S. households boast an income of $100,000 a year. With these dollars, the residents typically consume 40 percent more energy in heating, air conditioning, and electrical slaves than do people making $15,000 a year. The fuel efficiency of the mechanical slaves parked in their heated garages did not improve by even a gallon between 1986 and 2006. The richest world cohort in human history jets about the globe to shop, travel, or kill boredom. "The energy cost of their extensive air travel alone may prorate to more refined fuel per month than most families use in their cars per year," says Smil.

But this massive consumption of energy shouldn't be confused with quality of life. After looking at a variety of indicators, Smil made a surprising discovery: once individual consumption levels had reached seven barrels of oil equivalent a year, not much happiness was gained by burning more energy. In fact, energy consumption above 17 barrels yielded rapidly diminishing returns. Low infant mortality, a healthy diet, high life expectancy, and decent housing, say Smil, can all be achieved with energy spending three times less than what the average North American now throws away. "Insofar as political freedoms are concerned," he adds, "they have little to do with any increases of energy above the existential minima [1.5 barrels a day]; indeed some of the world's most repressive societies have high or even very high energy consumption."

Low-energy cultures have always understood these truths. Before oil fundamentally reengineered the American character, Thomas Paine, one of the nation's radical founders, noted, "'tis dearness only that gives everything its value." Excessive energy use has alienated the United States from the very ideals championed so elegantly by Paine: "When it shall be said in any country in the world, my poor are happy; neither ignorance nor distress is to be found among them; my jails are empty of prisoners, my streets of beggars; the aged are not in want, the taxes are not oppressive; the rational world is my friend. Because I am the friend of its happiness; when these things can be said, then may that country boast its constitution and its government." No American could believably make such a boast today.

High energy consumption nourishes highly narcissistic cultures. Since the 1970s, when domestic oil production peaked in the United States, the country has refused to recognize any real change in its energy fortunes. As noted by the essayist Daniel Altman, self-obsessed Americans tend to do what they want regardless of the consequences for other people. Some believe that they are entitled to superhuman wealth even when the country's dwindling oil supplies deliver nothing but debt. Many Americans consistently reject taxation that might serve future generations or the current good of their communities. Explains Altman, "In recent decades Americans have encountered far more inequality and far less social mobility than their parents. But narcissism leads these same Americans to reject redistributive tax systems."

Given his findings, and the fact that it would take five times more energy than the current global supply to extend these profligate energy habits to the rest of the world, Vaclav Smil proposes something other than this "utterly impossible option." He wrote that "we had plenty to gain earlier as we were moving along the energy escalator—but now the affluent world is within the realm of limited to grossly diminished returns." The libertarian pursuit of more and more energy must be replaced with a more ethical and conservative imperative: limiting consumption.

Most people think that this goal can be achieved through energy conservation or alternative energies, including giant windmills and industrial fields of solar panels. Smil doesn't agree. For starters, energy efficiencies almost always lead to greater consumption of energy in the form of bigger homes or faster cars. William Stanley Jevons, a Victorian economist, first described the phenomenon when contemplating England's ravenous consumption of coal. He noted that as blast furnaces improved and required less coal, industry responded by building more blast furnaces to make more cheap things, boosting total coal consumption. "It is wholly a confusion of ideas to suppose that the economical use of fuel is equivalent to a diminished consumption. The very contrary is the truth. As a rule, new modes of economy will lead to an increase of consumption according to a principle recognized in many parallel instances." The Jevons paradox explains why the average North American household now spends more of its electrical budget on TVs and computers than it does on clothes washers or dryers.

Renewables, by definition, harvest less energy than do densely packed fossil fuels. Like unconventional hydrocarbons such as bitumen and shale gas, they require intensive industrial farming. They also create greater landscape disturbances. To replace a thousand-megawatt coal-fired plant sitting on 1.5 square miles of land with solar panels would require a small city-sized area of 19 square miles. To achieve the same energy gains with wind could take an area three times bigger than the solar plantation. In fact, meeting one-third of the world's current energy needs with wind power would require approximately 13 million towers spaced half a mile apart occupying 3 million square miles: approximately 5 percent of the world's total land mass. This megaproject would cost at least $15 trillion. Harvesting crops to produce biofuels that could replace oil as a transportation fuel would require the domination of 50 percent of the earth's solar plant energy. To produce biofuels from algae ponds would require an area the size of Iowa just to power the United States. "Only an utterly biologically illiterate mind could recommend such action," says Smil. All currently proposed wind, solar, thermal, and biomass projects still add up to less energy than the current global consumption of 15 terawatts. (The terawatt is an ungainly energy measurement: it is equal to 5 billion barrels of oil, or the power of a brief lightning strike.) That doesn't mean we shouldn't pursue better efficiencies and invest in durable photovoltaics, but we can't get far enough without also restraining energy use.

A group of American and Russian physicists reached similar conclusions in 2008 in a study published in *Ecological Complexity*. Their article makes some damning observations about the potential of renewables and future energy policy. Authors Anastassia Makarieva, Victor Gorshkov, and Bai-Lian Li calculated that humans consumed, in 2005, about 15 terawatts of power, largely from fossil fuels. The world's forests, rivers, oceans, winds, and plants once circulated energy equivalent to 100 terawatts, nearly seven times what humans now command. But thanks to the brute force of billions of inanimate slaves, humans have now destroyed 60 percent of the globe's biological systems and their energy flows. As a consequence, "global regulatory biotic power" has been sorely diminished, leaving the world's ecosystems with only 40 terawatts of power.

Massive renewable projects would severely draw down upon this remaining system, if not sabotage it. Forests control wind or atmospheric circulation on land largely by pumping moisture into the air. This critical service moves water from the land to the oceans. Massive wind farms can impede this movement of water. The group calculated that a wind farm composed of 9 million towers occupying 23 percent of North America would affect storm activity in the Atlantic and also change local weather. "The use of wind power is equivalent to deforestation," explained the physicists. As a consequence, wind could never deliver more than 5 percent of human energy needs "and will never be able to compete in power with the existing hydrological dams." Yet energy experts now forecast an impossible fantasy: a world in which humans will wield 44 terawatts of power, with 10 percent of that coming from industrial wind farms.

Similar limits apply to solar power. "Any large-scale consumption of solar energy, currently perceived as harmless and environmentally friendly, will have a catastrophic impact on the resilience of these critically important life-supporting processes," noted the American and Russian scientists. Their findings indicate that renewables could at best provide for no more than one-tenth of human energy consumption without destroying the delivery of critical natural energy flows upon which all life now depends. Business as usual using renewables will simply aggravate a "dangerous situation." Like Smil, the authors conclude that stability can be achieved only by reducing human pressure on global energy flows and lowering global energy consumption.

Lowering energy demand is a radical prescription. It is as revolutionary as the abolition of U.S. slavery or the redistribution of land in nineteenth-century Russia, says Vaclav Smil. But our health, our freedom, and our humanity depend on a moral reassessment of mastery and slavery in all energy relationships. Our debilitating servitude to the concentrating forces of fossil fuels has just one proper solution: a radical decentralization and relocalizing of energy spending combined with a systematic reduction of the number of inanimate slaves in our households and places of work.

Ivan Illich, a Catholic theologian and medieval scholar, arrived at the same conclusion in 1974. In *Energy and Equity*, one of the most brilliant essays ever written on the subject, the maverick thinker argued that high energy consumption degraded human relationships as inevitably as it destroyed watersheds, mountains, and forests. Societies that overindulge in energy consumption, he warned, ultimately lose their freedom, their resilience, and their independence. Low-energy societies protect the oldest freedom—the possibility of walking—and offer greater choice and participation in everyday life, argued Illich. They build on a human scale and place all the needs of daily life within the realm of a walker.

To Illich, high-energy societies such as the United States and Saudi Arabia had crossed a threshold into brutal mechanical slavery. They had become dependent on a technocracy, a class of petroleum and nuclear engineers who dictated the terms of existence in purely energy terms. As societies consume more energy, with more energy slaves to support, Illich believed, they become more complex and totalitarian. In high-energy societies, "man is born into perpetual dependence on slaves which he must painfully learn to master. If he does not employ prisoners, then he needs machines to do most of his work," wrote Illich. "According to this doctrine, the well-being of a society can be measured by the number of years its members have gone to school and by the number of energy slaves they have thereby learned to command. This belief is common to the conflicting economic ideologies now in vogue. It is threatened by the obvious inequity, harriedness, and impotence that appear everywhere once the voracious hordes of energy slaves outnumber people by a certain proportion. The energy crisis focuses concern on the scarcity of fodder for these slaves. I prefer to ask whether free men need them."

Illich had no illusions about energy and equity. The more a society consumed, the less equity it possessed. "Even if nonpolluting power were feasible and abundant, the use of energy on a massive scale acts on society like a drug that is physically harmless but psychically enslaving. A community can choose between Methadone and 'cold turkey'—between maintaining its addiction to alien energy and kicking it in painful cramps—but no society can have a population that is hooked on progressively larger numbers of energy slaves and whose members are also autonomously active." The poor must abandon North American energy fantasies, the theologian believed, while the rich must "recognize their vested interest as a ghastly liability. Both must reject the fatal image of man the slaveholder currently promoted by an ideologically stimulated hunger for more energy."

Illich thought that moving from a culture of high energy spending to one of low-energy living would require three steps. First, society must set some kind of limit on the per capita use of energy. Second, the community must discuss what living on less energy really means and what individual threshold would work locally. Illich thought the energy needed might fall somewhere between the power of a horse and that commanded by a Volkswagen. Last, each community must choose between the bicycle and the car, between a "postindustrial, labor-intensive, low-energy and high-equity economy" and the "escalation of capital-intensive institutional growth" that would lead to a "hyperindustrial Armageddon."

From 1937 to 1941, the brilliant sociologist Pitirim Sorokin, who survived both czarist and Bolshevik oppression before escaping to the United States, published a long and provocative treatise on the rise and fall of human cultures. In *Social and Cultural Dynamics*, Sorokin identified energy as the driver of culture; its abundance or scarcity could define the character of civilizations. The Christian anarchist also saw a clear pattern in the nonlinear life of civilizations. According to Sorokin, civilizations generally pulsated between two distinct poles: one was dominated by spiritual enlightenment; the other pursued material pleasures. While "ideational" cultures spurned money and wealth, "sensate" cultures actively sought power and enjoyment. One existed on low levels of energy, while the other mushroomed on the temporary surpluses created by agricultural innovation, new sources of slavery, and the advent of fossil fuels. Some of the world's most interesting, "idealistic" cultures accommodated a bit of both.

To Sorokin, the instability and imbalance created by rapid change in sensate cultures rarely supported its mission of increasing happiness. The "needs and aims" of a sensate culture, he wrote, "are mainly physical, and maximum satisfaction is sought of these needs. The method of realizing them is not that of modification with the human individuals composing the culture, but of a modification or exploitation of the external world." Moreover, sensate culture unceasingly "tries to be 'progressive, dynamic'" by seeking "new empirical values"; it "values the latest fashion instead of the old-time consecrated tradition. It tears down the building just erected to replace it by a new one. It puts a premium upon anything swift, fast, dynamic, modern, 'up to the last minute' and even beyond it. Hence its feverish tempi of change, its insatiable lust for change, its never-resting Becoming."

Sorokin noted that high-energy cultures characterized by rapid technological and scientific advances tended to be more warlike. To prove this point, he even tallied the dead from the main European conflicts between the twelfth and the twentieth centuries. "The medieval centuries were predominantly monarchical and autocratic, illiterate, and possessed of very few scientific discoveries and technological inventions, yet the level of war was low. In the subsequent centuries, beginning with the thirteenth, discoveries and inventions, literacy and education grew steadily, especially in the nineteenth and twentieth centuries, yet wars constantly increased from the twelfth to the eighteenth century, and in the twentieth reached a magnitude probably unequaled in the entire history of the human race."

Sensate cultures, with their high-carbon materialism, infect everyone with a particular mindset. "Sensate minds emphatically disbelieve the power of love, sacrifice, friendship, co-operation, the call of duty, unselfish search for truth, goodness, and beauty." Such minds are prone to believe in "the power of the struggle for existence, selfish interests, egoistic competition, hate, the fighting instinct, sex drives, the instinct of death and destruction, all-powerful economic factors, rude coercion and other negativistic forces."

Just as sensate Roman civilization eventually gave way to the ideational Christianity of the medieval ages, so every ruling culture contains the seeds of its own destruction, said Sorokin. Slaveholders possessed this peculiar one-sidedness just as oil executives would. "Each supersystem becomes increasingly sterile and progressively hinders the emergence of a new and vital supersystem representing an aspect of reality largely neglected during the domination of its predecessor. Such a situation presents, as it were, an ultimatum to the society and culture in question: they are forced either to replace the exhausted supersystem with a creative one or else to become stagnant and fossilized." A change in the availability of energy marks each change, said Sorokin, who died in 1968. He had no doubt that our hydrocarbon-based culture had become overripe, narcissistic, and decadent and would fail, like all civilizations, in unpredictable ways.

The rising unhappiness of Americans over the last century has been well documented by a number of scholars. In his famous study *Bowling Alone*, political scientist Robert Putnam noted that mobility and obsession with material things had come with a high civic price. Based on 500,000 interviews, Putman found that Americans had lost the value of friendship (what oil-funded academics call "social capital") and conviviality. They rarely talked to neighbors or attended public meetings, and they belonged to fewer social clubs. They locked their doors and signed fewer petitions. They invited friends and relatives over for dinner less often. They spent more time glued to screens or alone in their cars, commuting. Based on Putnam's data, collected between 1975 and 1998, the places with the lowest "social capital" ratings included the petrostates of Texas, Oklahoma, and Louisiana.

Professor of social work Brené Brown at the University of Houston has echoed these findings. Brown studies vulnerability, courage, shame, and how people connect. (Perhaps only petroleum wealth could create an academic field called "vulnerability research.") Her studies on the importance of connection eventually led Brown to a spiritual crisis. The scientist in her was forced to recognize the nature of being human. In 2009, Brown gave an inspiring TED talk called "The Power of Vulnerability." Although she didn't use the words *petroleum* or *energy*, her speech catalogued the precipitous cost of high-energy consumption on human emotions. "We are the most in debt, obese, addicted and medicated adult cohort in U.S. history," Brown declared. She identified the numbness that comes from the inability to experience genuine shame, grief, and disappointment. "You cannot selectively numb. So when we numb those, we numb joy, we numb gratitude, we numb happiness. And then we are miserable, and we are looking for purpose and meaning, and then we feel vulnerable, so then we have a couple of beers and a banana nut muffin. And it becomes this dangerous cycle... Religion has gone from a belief in faith and mystery to certainty. I'm right, you're wrong. Shut up. That's it... This is what politics looks like today. There's no discourse anymore. There's no conversation. There's just blame. You know how blame is described in the research? A way to discharge pain and discomfort."

Earl Cook, the renowned Texas A&M geologist, wrote a perceptive critique of high-energy living. "Affluence, freedom, and waste," he declared, "are the three musketeers of the growth society." Cook identified two big barriers to change: entrenched elites and the myths that support the customs of high-energy living. "The service layers of high energy society, especially those highly paid who enjoy the status of oracles, such as economic advisers to the government, medical specialists in the diseases of affluence, media commentators, professional athletes, counselors of the lustlorn, and pedagogic protectors of the cultural heritage, wince at the suggestion of a re-examination of their social usefulness that may be called for when scarcity replaces affluence." The myths that bind high-energy societies and blind their citizens form a wall as obstinate and thick as the divide that separated Berlin: "that the market economy allocates even scarce resources more beneficially than can government; that the growth state is necessary for social well-being; that technology is exempt from the law of diminishing returns; that both government and the press inform the citizen so that he exercises his franchise in a more intelligent fashion; and that although man may not live by bread alone, he makes decisions as if he did." To end the upheaval, revolutions, and wars that had marked the ascent of high energy consumption, Cook wrote, "we must abandon efficiency of production as a social goal and replace it by efficiency of consumption. We need to favor measures of quality of achievement over measures of quantity. And we must stop confusing momentum with progress and growth with goodness."

Nearly fifty years ago, U.S. sociologist Fred Cottrell offered some sobering reflections on life in a time of energy descent. A world experiencing diminishing supplies, he noted, would have to make sacrifices. "Choices must be made between extravaganzas like going to the moon and guaranteeing pure water," he wrote. The growth of centralized government will stop and wither. Moreover, the energy descent will unleash endemic conflict, and "those whose greatest advantage is in the market place will struggle to protect it." Government, once concerned about managing the flow of energy, will now be pressed to deal with rising prices and the flow of food. People's values will change in response to diminished energy incomes in unpredictable ways. "Just as the removal of colonial dominance has led to a resurgence of old cultures and old values in many parts of the world," said Cottrell, "the development of different energy systems that offer different alternatives than those provided by the omnicompetent giant corporations, may demonstrate that many Americans prefer a lifestyle different in significant ways from those that now prevail, and can be attained with the use of far less energy than is now used."

Toward the close of 2011, an American libertarian entrepreneur and a French socialist banker complained bitterly about loss of faith in the future. The malcontents appeared in separate stories in *The New Yorker.* Separated by the Atlantic Ocean and several issues of the magazine, the businessmen sound like long-lost twins. Each was wrestling with the decline caused in his country by inanimate slaves running on expensive fossil fuels.

One of the men was Peter Thiel, the billionaire founder of PayPal, who has obsessed for years about the dismal state of the U.S. economy and the obliviousness of its elites. In the *New Yorker* profile, he speculated about why the United States hasn't made the expected big breakthroughs in life extension and artificial intelligence. The country seems to have lost its edge, said the digital guru. Even its science fiction no longer offers high-energy visions of flying cities or underwater homes; instead it portrays galaxies run by fundamentalist terrorists armed with technologies that fall apart. The United States no longer seemed interested in technological marvels, observed Theil: "You have dizzying change where there's no progress... Everyone knows things are rotten." The billionaire said that energy and food are very "politically linked" and that he studiously avoids politics.

Matthieu Pigasse, part owner of *Le Monde,* expressed similar concerns about the loss of greatness in France. Pigasse diagnosed all of Europe as depressed, both economically and emotionally. Five years of no economic growth and no rise in consumption, wages, or investment had yielded something new and unexpected, he said: "political instability." Just about every government in the European Union, from Italy to Belgium, was in disarray, leaderless, or fantastically in debt. "I think we are at a crossroads, and everybody knows that, but nobody wants to confront the reality," Pigasse believes. He hopes that bigger government and a greater economic union might somehow offer a solution.

All energy issues are moral ones. The Christian philosopher C.S. Lewis grasped this fundamental truth with a gentle fierceness. In *The Abolition of Man,* Lewis wrote, "Each new power won *by* man is a power *over* man as well. Each advance leaves him weaker as well as stronger. In every victory, besides being the general who triumphs, he is also the prisoner who follows the triumphal car." Lewis compared the modern fascination with power to an Irish folktale: when a fellow discovered that a certain kind of woodstove reduced his fuel bill by half, he ordered a second stove, in order to warm his house with no fuel at all. "It is the magician's bargain: give up our soul, get power in return. But once our souls, that is, our selves, have been given up, the power thus conferred will not belong to us. We shall in fact be the slaves and puppets of that to which we have given our souls."

The pursuit of right livelihood now leads all thinking people in one direction only: the consumption of less oil and less industrially packaged energy. In chasing happiness with energy, we have lost it. Emancipation will not proceed in a manner that is predictable, rational, or linear. It may begin, however, with a meditation on the words of Leo Tolstoy: "Energy rests upon love; and come as it will, there's no forcing it."

13

Japan and the Fragility of the Petroleum Age

. . . . .

"The classical struggle between order and disorder, between angels and evils, is still with us."

HOWARD ODUM, *Environment, Power and Society*, 1971

THE GREAT Sendai earthquake, tsunami, and nuclear power meltdown of 2011 will go down in the annals as just three in the series of unfortunate events illuminating the world's downward energy spiral. As the writer Kenzaburō Ōe painfully noted, "Japanese history has entered a new phase." Although the media fixated on the ever-expanding risks of nuclear power, what the quake really laid bare were the fragilities of an oil-fueled economy that has peaked and lost its momentum. The tsunami and the Fukushima meltdown swept away whatever delusions the nation's elite had about a graceful decline.

The slow death of what was once the world's number-three economy, with a GDP greater than that of France and Britain, reads like a tragic twentieth-century love story. Two decades ago Japan boasted 14 percent of the world's economic output; today its share is less than 8 percent and falling. Even before the disasters of 2011, the Japanese government had predicted that as many as two thousand rural communities must cease to exist in the next decade, due to persistent economic decline in "the shrinking regions." As export-oriented industries close their doors, villages stand forlorn and empty. Tokyo, the world's first megacity, has stopped growing. The landscapes that oil emptied to forge unsustainable urban centers have now been reclaimed by monkeys and boars. "There is no historic precedent for such a rapid move from globally admired economic superpower to floundering society enmeshed in a prolonged socioeconomic crisis," notes the energy historian Vaclav Smil. Having no indigenous oil sources, Japan is an energy orphan and, therefore, a global everyman.

The Sendai earthquake was a novel power event. It generated about 476 megatons of energy, 600 million times more energy than the bomb that leveled Hiroshima. The energy released by the quake shifted one of the main islands of Japan by 8 feet and lowered the coastline by 3. It also moved the earth's axis several inches and shortened the length of Earth's day. The tectonic rattling created giant waves that swept more than 20,000 people and 8 million tons of debris out to sea. It destroyed most of the energy infrastructure of northeastern Honshu, including harbors, airports, and refineries. Even Tokyo's iconic electronic blackboards temporarily stopped flickering.

But this sudden devastation was no more chilling than the slow-moving Godzilla of petroleum. For nearly fifty years, Japan has gobbled oil in order to build a highly complex consumer culture. It remains the world's third-largest importer of petroleum, at 4 million barrels a day. (That's double Canada's daily tar sands production.) Japan meets nearly 50 percent of its primary energy needs with oil, and oil accounts for nearly a third of the country's imports in value. The lifestyle of the average Japanese consumer now demands eighteen barrels a year. About 90 percent of those barrels are filled with oil from the Middle East.

Japan, like every nation, dances with petroleum in its own way. In the United States, oil energized the county's pioneering spirit and created an adolescent business culture with no memory. In the Middle East, oil allowed poor sultans to green deserts and fund religious fanaticism. In Japan, it enabled a thousand-year-old warrior culture, a Chosen People, to "pave and build" with belligerent intent. Long buffeted by monsoons, floods, typhoons, earthquakes, and tsunamis, the Japanese sought a way to tame Nature's costly outbursts.

In a remarkable book on the failure of modern Japan, Alex Kerr, a longtime resident of the shrinking island, describes how the country married its fear of being "the Archipelago of Disaster" with an industrial psychology of "Total Dedication." It sweetened the match, Kerr writes, "with a dowry in the form of rich proceeds to politicians and bureaucrats." "Government-paid propaganda singing the praises of dam and road builders" glorified the process. The result, says Kerr, was "an assault on the landscape that verges on mania," "an unstoppable extremism" reminiscent of Japan's military buildup before World War II. After failing to establish its place in the sun during the 1930s and 1940s with military expansion and an army of 10 million Chinese slaves, Japan regrouped. "All the forces of the modern state" were "made to focus on eradicating nature's threats." Oil provided the means for what economists dubbed "the Japanese miracle."

Ancient Japan, of course, employed slaves. As many as 10 percent of its people served as slaves for temples, shrines, and public officials or as chattel for wealthy farmers. (They had the worth of a good cow.) Before the petroleum age, the Japanese archipelago ran on rice, peasants, and human labor. It boasted pure mountain streams and unspoiled shores. The majority of people lived in rural villages. Kyoto, an Asian Venice, was so beautiful that the Americans dared not bomb it during WWII. Prior to the Industrial Revolution, the island's population never climbed above 35 million.

But with more energy, Japan changed its metabolism and launched a demographic revolution. In 1905, a population of 47 million, fired by coal, defeated Russia. In 1931, a population of 67 million, emboldened by small amounts of petroleum, invaded China for more resources. Having become addicted to fossil fuels, Japan now needed a larger industrial farm base to feed its people. In 1941, a population of 72 million challenged the United States, the world's mightiest petrostate. After the war, with the American gift of cheap oil, a defeated population rebuilt its landscape and became an industrial superpower. On petroleum, Japan boosted its numbers to 120 million.

Within two to three decades, oil had concentrated 79 million Japanese—70 percent of the population—into cramped homes in 209 complex urban centers. According to Kerr, these new cities offered "an apocalyptic expanse of aluminum, Hitachi signs, roof boxes, billboards, telephone wires, vending machines, granite pavement, flashing lights, plastic, and pachinko." Whenever the air pollution got unbearable, people just wore gas masks to work. But cheap oil did more than concentrate power and people. It changed the nation's economic thinking, too. Oil allowed Japan, a nation with few resources, to import raw goods and turn them into electronic gadgets and cars for global export. The more oil Japan consumed, the higher its GDP rose. At one point it was the second highest in the world.

Thanks to the surplus profits generated by its oil miracle, Japan built high-speed trains, factories, bridges, and millions of cars. It became a "construction state." Aluminum replaced bamboo and concrete replaced wood. To show off its newfound energy prowess, Japan hosted the Olympics in 1964. The event announced Japan's rebirth to the world.

Oil also gave Japan the power to reengineer the country's natural energy flows. The country clear-cut its broadleaf forests and replaced them with fast-growing cedars. It dammed its rivers and turned streams into concrete-lined canals. It leveled mountains. Japan's shores were lined with giant tetrapods, ungainly concrete blocks meant to block waves and prevent erosion (these oil monuments only accelerated the process). Rural communities were transformed into urban waste dumps. Oil encouraged the nation's elites to dream like gods. Kōnosuke Matsushita, founder of Panasonic, once proposed a two-hundred-year energy project that would excavate 20 percent of Japan's mountains to create a fifth island the size of Shikoku.

Oil changed the Japanese diet as well. Out went local rice, vegetables, and fish; in came imported meat, fat, and grains. "The excessive consumption of cereals has diminished with an increasing consumption of meat, milk and dairy products," bragged the government. Today, Japan is the least food self-sufficient of any industrial nation.

After the global oil shocks of the 1970s, Japan built more than fifty nuclear facilities on the Ring of Fire. Scientists believed this seemingly inexhaustible source of energy would finally solve the resource-poor nation's energy dilemmas.

The highly subsidized atomic industry provided 30 percent of the nation's power and its professional engineers and spin doctors dominated the politics of the nation's ten regional electrical monopolies. (That investment, combined with a real estate bubble, explains why Japan now has the largest national debt of any other economic superpower except the United States.)

But taming atomic particles to lessen Japan's dependence on oil proved more difficult than imagined. The Mihama reactor leaked in 1991. A fire and an explosion exposed hundreds to radiation in 1997 in Tokaimura. A steam explosion killed four workers in 2004, again in Mihama. The Tokyo Electric Power Company routinely falsified safety audits. To convince ordinary people that nuclear reactors were safe, the scientific establishment and the Japanese state built swimming pools, golf courses, and movie theaters close to the reactors. The industry targeted women and children with nuclear propaganda. "The companies believed that mothers were key decision makers in the family," explained anthropologist Noriya Sumihara, and if women "felt the plants were relatively safe, then men would, too." The cartoon characters Uranium Boy and Little Pluto reassured children, "No need to worry too much!"

Oil's miraculous powers too had waned; they peaked about fifteen years ago. One of the world's most energy-efficient nations suddenly found it couldn't squeeze much more out of a barrel. As the population aged (a fifth of the Japanese population is over sixty-five) and the cost of oil imports grew dearer, the economy stagnated. In 2009, Japan's GDP shrank by 15 percent and oil consumption declined by nearly a million barrels a day. James Howard Kunstler, a perceptive energy critic, accurately described the phenomenon in his

2011 forecast: "A decline in the primary energy resource used by an industrial society portends a decline in living standards, which can be expressed in an economy, for instance, by people having less money, or by people having lots of money that is increasingly worthless."

The *New York Times* has dubbed this process "Japanification." Nearly 40 percent of Japan's population is unemployed or underemployed. The Japanese middle class no longer fly to Honolulu in organized groups to buy Gucci bags. They've traded in their German cars for cheaper Japanese models. The national government spends a quarter of all taxes on paying interest on the national energy debt. And today, as in all industrial kingdoms, the country's fertility rates have declined. Japan's is a shrinking society in which old people outnumber workers. Every year Japan loses the equivalent of 490,000 people. By 2050, the country's population could fall below 100 million. "There is, of course, no precedent in history for a population where the numbers of octo- and nonagenarians would surpass those of children," writes Vaclav Smil. "By 2050 Japan could have nearly 5 million people in their 90s and there could be more than half a million of centenarians, some 90% of them women."

Faced with revolution or stagnation, it's unclear which road the Japanese will choose. But entropy has set in. The country's atomic behemoth sucked money away from renewables (Japan used to be a solar leader) and created an elite bureaucracy of liars and thugs. Japan's 2010 Basic Energy Plan called for the construction of nine more nuclear plants and promised a new economic miracle: a global nuclear export industry. But the 2011 earthquake and the Fukushima meltdown ended that crazy dream. With the world's oldest population and a baby-booming generation shaped and conditioned by the illusions of petroleum, Japan may no longer have the workforce to connect with reality again.

The Japanese people could take advantage of the situation to demand reform. The country could fall back on traditions of resilience. The Zen masters knew about the impermanence of earthly things and the inexhaustible beauty of nature. They did not fear impermanence, nor did they seek to control it. They regarded Nature's force and unpredictability as daily reminders to live well, appreciate life, and do good work. The twelfth-century philosopher Kamo no Chōmei once advised, "If you have to go anywhere, go on your own feet. It may be trying, but not so much so as the bother of horses and carriages. Every one with a body has two servants, his hands and feet, and they will serve his will exactly."

For some citizens, powering down has become a passion. Twelve years ago, Yasuyuki Fujimura, an engineer and inventor, started the "non-electric movement." Today it is booming. Fujimura believed that Japan's demand for electrical slaves had supported a corrupt nuclear empire, a high suicide rate, and much general unhappiness. He calculated that electric rice cookers alone consumed enough energy to power 2.4 nuclear reactors every year. The inventor has developed some 1,000 household tools that don't require electricity, including a dust sweeper and a solar rice cooker. On his website Fujimura explains, "To gain one thing means to lose another." He says turning off switches will foster independence and spiritual well-being.

If Japan's leaders ever take up walking, they will encounter the hundreds of stone markers that dot the island's coastlines. The tsunami washed away some but revealed others. Some of these monuments are six hundred years old and stand ten feet tall. Many bear the high-water marks of previous tsunamis. Others contain warnings from the nation's low-energy ancestors. One says, "High dwellings ensure the peace and happiness of our descendants. Remember the calamity of the great tsunamis. Do not build any homes below this point." Another reads, "No matter how many years may pass, do not forget this warning." But oil-confident Japan put its faith in tetrapods and petroleum's inanimate slaves. It ignored the stones.

Epilogue

. . . . .

AFTER THE "dark night of history" that followed the collapse of the Roman Empire, new flows of energy slowly recharged rural Europe. They often sprang up in unlikely places, on the margins of things. Unlike Rome's slaveholder elites, who had put material pursuits first, these new low-energy communities resolutely heralded a simple life. One of the most famous began at Monte Cassino outside Rome in the sixth century. Here Benedict of Nursia, the son of a wealthy Roman family, founded what would become the Benedictine order.

Benedict not only created a new community based on prayer, learning, and human labor but wrote a constitution for religious orders. His *Rule* was a masterly plea for discipline and humility. It prized equality, chastity, service, and sharing. "Let not one of noble birth be put before him that was formerly a slave," said Benedict. His rulebook praised what Rome had abandoned—adaptability and flexibility—and elevated the idea of manual labor from the degradation of institutional slavery. A virtuous person served God by living and working in a community that provided its own sovereign energy. Self-reliance was paramount. Explained the rule book: "The monastery ought if possible be so constituted that all things necessary such as water, a mill, a garden, and the various crafts might be contained within it."

By design, Benedictine communities, whether monasteries, priories, or abbeys, were autonomous, and they generally operated on a small scale. The largest boasted nine hundred clerics. Most sheltered no more than a dozen or two voluntary members. What distinguished the order, in the eyes of U.S. social critic Lewis Mumford, was its acceptance of work "not as a slave's curse but as part of a free man's moral commitment." And if "the Benedictine motto is 'to labor is to pray,'" Mumford added, "this meant that offices of ritual and work had at last become transferable and interchangeable; yet each part of life was directed toward a more exalted destination." In simple terms, Benedict celebrated human energy for higher ends. Many members of the order went on to achieve immense fame. The Benedictine nun Hildegard of Bingen, for example, wrote on medicine, theology, and music and served as an adviser to popes and kings. "Love life and use your five senses correctly," she counseled.

Benedictines devoted most of their energy surpluses to the arts. In their well-ordered fields, orchards, and buildings, the monks worked only five hours a day with their hands. They spent the remainder praying, reading, copying manuscripts, perfecting crafts, or conversing. Each monk worked at a variety of ever-changing tasks and shared in the fruits of these labors. In old age, monks received good care and other privileges. After the horrors of slavery, they showed that a Christian community could give every person a chance to express his or her full qualities. As Mumford explained, "Rewarding work [the Benedictines] kept for themselves: manuscript copying, illumination, carving. Unrewarding work they turned over to the machine: grinding, pounding, sawing. In that original discrimination they showed their intellectual superiority to many of our own contemporaries who seek to transfer both forms of work to the machine, even if the resultant life proves to be mindless and meaningless."

The Benedict monasteries flourished for a long time. By the eleventh century, more than fifteen thousand self-supporting Benedictine villages populated Germany, England, and northern Europe. Sited on streams, these innovative religious centers became perhaps the most successful business enterprises on earth. The communities grew their own food and often sold the surplus at less than market prices, to discourage greed. They shared crop knowledge and medicines with local peasants. They welcomed visitors and cared for the sick. Every monk took turns cooking and serving food at the dinner table. They read aloud to each other. They painted shrines, made glass, enameled chalices, and produced organ pipes. All in all, they behaved like the best of the world's agrarian families.

As Lewis Mumford noted, the monks initially developed specific tools to free themselves from repetitive tasks. They adopted the horse-powered treadmill, then the water mill and the windmill. Beginning in the tenth century, the Benedictines and other orders helped to spread the use of water mills across Europe and England. But as owners of water mills and their energy surpluses, the monasteries eventually amassed much status and authority. And the labor-saving technology gave religious houses another problem: a surplus of coin. The mills provided such a rich and constant source of revenue that by the eleventh century, critics were chastising the Benedictine order for its avarice and opulent living. The greed of monks dominated twelfth-century religious debates. Protestants went so far as to portray monks as evil creatures shat out by devils.

Not all monasteries used their mills to expand their waistlines and their authority. Many smaller orders, such as the Augustinians, milled goods at a fair exchange. The Cistercians built roads, dug canals, and constructed furnaces to forge metal. The Franciscans and the Dominicans lived among the poor. Unwittingly, though, these industrious orders helped to deforest much of northern Europe. The Great Deforestation created a wood scarcity that invited the English to either burn coal or freeze to death. The consumption of hydrocarbons eventually secularized the continent, disconnecting Europeans from self-sufficient work and from any ethical understanding of how to use energy well.

The rise and fall of the Benedictines reads like any energy fable. In their early, prayerful work, the monks used energy with grace and balance. They demonstrated that scarcity did not mean poverty; that abundance did not mean affluence; that progress did not mean momentum; that goodness did not mean growth. They prized quality of achievement and did not overvalue efficiency. With loving hands, they showed that a revolution in mind and heart could transform the wasteful energy orgies of the Roman Empire into a life that was lean and rewarding.

Alasdair MacIntyre, the British moral philosopher and author of *After Virtue*, thinks the world is now waiting not for Godot but for more St. Benedicts, albeit ones suited to the present day. What we urgently need, says MacIntyre, a former Marxist turned Catholic, are individuals dedicated to renewing civility and community on a worldwide scale. MacIntyre believes that many contemporary men and women want to participate in a different moral narrative than servitude to a petroleum order. The citizens who abandoned Rome for the haven of their agricultural estates were fleeing the hubris of concentrated energy and its attendant slavery. Today, resistance to the abuse of energy can begin, literally, with a walk, says MacIntyre, and each walk takes place on the grounds of hope.

And indeed, a haphazard and improbable emancipation movement has begun to take shape. Around the world, families and groups of individuals are walking away in ever-growing numbers from petroleum and the inanimate slave culture of frantic consumption. They are exchanging quantity for quality and relearning the practical arts. Those seeking liberty eat slowly, travel locally, plant gardens, work ethically, build communities, share tools, and eschew bigness in economic and political life. Above all, they are relearning what it means to live within their means, with grace. Like the Greeks long before them, these new abolitionists have come to understand that the indiscriminate spending of energy is mere Promethean hubris. Unqualified power diminishes life, the only true wealth we share. By burying the chains, we can find a new livelihood and an old freedom.

Sources

. . . . .

GENERAL WORKS

Blair, John. *The Control of Oil*. New York: Pantheon, 1976.

Chesterton, G.K. *The Outline of Sanity*. 1926; Norfolk, VA: IHS Press, 2001.

Cook, Earl. *Man, Energy, Society*, San Francisco: W.H. Freeman, 1976.

Cottrell, Fred. *Energy and Society: The Relation between Energy, Social Changes, and Economic Development*. New York: McGraw-Hill, 1955.

Crosby, Alfred W. *Children of the Sun: A History of Humanity's Unappeasable Appetite for Energy*. New York: Norton, 2006.

Debeir, Jean-Claude, Jean-Claude Deléage, and Daniel Hémery. *In the Servitude of Power: Energy and Civilisation through the Ages*. London: Zed Books, 1991.

Flipo, Fabrice. "Energy: Prometheus bound or unbound? A conceptual approach." *S.A.P.I.EN.S.* 1, no. 2 (2008).

Hall, Charles, and Kent Klitgaard. *Energy and the Wealth of Nations: Understanding the Biophysical Economy*. New York: Springer, 2012.

Illich, Ivan. *Energy and Equity*. London: Calder & Boyars, 1974.

Kohr, Léopold. *The Breakdown of Nations*. 1957; New York: E.P. Dutton, 1978.

Olien, Roger, and Diana Davids Olien. *Oil and Ideology: The Cultural Creation of the American Petroleum Industry*. Chapel Hill: University of North Carolina Press, 2000.

Smil, Vaclav. *Energy, Food, Environment: Realities, Myths, Options*. Oxford: Clarendon, 1986.

Smil, Vaclav. *Energy in Nature and Society: General Energetics of Complex Systems*. Cambridge, MA: MIT Press, 2008.

Smil, Vaclav. *Transforming the Twentieth Century: Technical Innovations and Their Consequences*. Oxford: Oxford University Press, 2006.
Stein, Richard. *Architecture and Energy*. Garden City, NY: Anchor, 1977.
Tainter, Joseph, and Tadeusz Patzek. *Drilling Down: The Gulf Oil Debacle and Our Energy Dilemma*. New York: Springer, 2011.
Tugendhat, Christopher. *Oil: The Biggest Business*. New York: G.P. Putnam, 1968.
Wagner, Robert. *Moby-Dick and the Mythology of Oil: An Admonition for the Petroleum Age*. Charleston, SC, 2010.
Wallace, Thomas. *Wealth, Energy, and Human Values: The Dynamics of Decaying Civilizations from Ancient Greece to America*. Bloomington, IN: AuthorHouse, 2009.
Weissenbacher, Manfred. *Sources of Power: How Energy Forges Human History*. 2 vols. Santa Barbara, CA: Praeger, 2009.
Yergin, Daniel. *The Prize: The Epic Quest for Oil, Money & Power*. 2001; New York: Free Press, 2008.

CHAPTER ONE

Angela, Alberto. *A Day in the Life of Ancient Rome: Daily Life, Mysteries and Curiosities*. Translated by Gregory Conti. New York: Europa Editions, 2009.
Bardi, Ugo. "Joseph Tainter: Talking about collapse." *Cassandra's Legacy* (blog), March 10, 2011. At cassandralegacy.blogspot.ca/2011/03/joseph-tainter-talking-about-collapse.html.
Bradley, K.R. *Slaves and Masters in the Roman Empire: A Study in Social Control*. 1984; New York: Oxford University Press, 1987.
Du Bois, W.E.B. *The Negro*. 1915; Forgotten Books, 2007.
Epictetus. *The Discourses*. At classics.mit.edu/Epictetus/discourses.html.
Finley, Moses. *Ancient Slavery and Modern Ideology*. 1980; Princeton, NJ: Marcus Weiner, 1998.
Genovese, Eugene. *The World the Slaveholders Made: Two Essays in Interpretation*. New York: Random House, 1969.
Holt, T.W. *The Right of American Slavery*. New York: Baker & Godwin, 1850.
Jordan, Winthrop D., ed. *Slavery and the American South: Essays and Commentaries*. Jackson: University Press of Mississippi, 2003.
Meadows, Donella H. "Thomas Jefferson and Donella Meadows, slave-owners." *The Donella Meadows Archive*. November 12, 1998. At sustainer.org/dhm_archive/index.php?display_article=vn770jeffersoned.
Montesquieu, Charles de Secondat, baron de. *The Spirit of Laws*. Translated by Thomas Nugent. 1752; Kitchener, ON: Batoche Books, 2001.
Mouhot, Jean-François. "Past connections and present similarities in slave ownership and fossil fuel usage." *Climate Change* 105, nos. 1–2 (2011): 329–355.
Oakes, James. *The Ruling Race: A History of American Slaveholders*. New York: Knopf, 1982.
Patterson, Orlando. *Slavery and Social Death: A Comparative Study*. Cambridge, MA: Harvard University Press, 1982.

Ramsay, James. *An Essay on the Treatment and Conversion of African Slaves in the British Sugar Colonies*. London, 1784.

Seneca, Lucius Annaeus. *Moral Essays*. Translated by John W. Basore. Vol. 1. London: Heinemann, 1928.

Tainter, Joseph, T.F.H. Allen, Amanda Little, and Thomas W. Hoekstra. "Resource transitions and energy gain: Contexts of organization." *Conservation Ecology* 7, no. 3 (2003).

Thomas, Hugh. *The Slave Trade: The Story of the Atlantic Slave Trade, 1440–1870*. New York: Simon & Schuster, 1997.

Tocqueville, Alexis de. *Democracy in America*. Translated by Henry Reeve. Vol. 1. 1838; Project Gutenberg, 2006.

Urbainczyk, Theresa. *Slave Revolts in Antiquity*. Berkeley: University of California Press, 2008.

van Loon, Hendrik Willem. "Ancient and mediaeval civilizations." In *Whither Mankind a Panorama of Modern Civilization*, edited by Charles Austin Beard. New York: Longmans, Green, 1928.

CHAPTER TWO

The Abolition Project. See abolition.e2bn.org.

Babbage, Charles. *On the Economy of Machinery and Manufactures*. 1833; London: J. Murray, 1846.

Boston Society for the Diffusion of Useful Knowledge. *American Library of Useful Knowledge*. Vol. 1. Boston: Stimpson and Clapp, 1831.

Carlyle, Thomas. "Signs of the times." *Edinburgh Review*, March–June 1829.

Carnegie, Andrew. *James Watt*. New York: Doubleday, Page, 1905.

Clarkson, Thomas. *An Essay on the Slavery and Commerce of the Human Species*. London, 1785.

Eltis, David. *The Rise of African Slavery in the Americas*. Cambridge: Cambridge University Press, 2000.

Farey, John. *A Treatise on the Steam Engine*. London: Longman, Rees, Orme, Brown, and Green, 1827.

Fourier, Charles. *The Phalanx, or, Journal of Social Science*. 1843; New York: Burt Franklin, 1967.

Freese, Barbara. *Coal: A Human History*. Cambridge, MA: Perseus, 2003.

Hochschild, Adam. *Bury the Chains: Prophets and Rebels in the Fight to Free an Empire's Slaves*. New York: Mariner, 2006.

Mann, Horace. *Slavery: Letters and Speeches*. Boston: B.B. Mussey, 1851.

Oakes, James. *The Ruling Race: A History of American Slaveholders*. New York: Knopf, 1982.

Smith, Adam. *An Inquiry into the Nature and Causes of the Wealth of Nations*. 2 vols. 1776; Oxford University Press, 1976; rpt. Indianapolis, IN: Liberty Fund, 1982.

Ubbelohde, A.R. *Man and Energy*. 1955. Revised ed.; Baltimore, MD: Penguin, 1963.

Whitehead, Alfred North. *Adventures of Ideas*. 1933; New York: Free Press, 1985.

CHAPTER THREE

Adams, Henry. *The Degradation of the Democratic Dogma*. New York: Macmillan, 1919.

Adams, Henry. *The Education of Henry Adams*. 1918; New York: Library of America, 2010.

Armentano, D.T. "The petroleum industry: A historical study in power." *Cato Journal* 1, no. 1 (1981).

Carson, Rachel. *Silent Spring*. Boston: Houghton Mifflin, 1962.

Co, Annie. "'Breed out the Unfit and Breed in the Fit': Irving Fisher, economics and the science of heredity." *American Journal of Economics and Sociology* 64, no. 3 (2005): 793–826.

Denny, Ludwell. *We Fight for Oil*. New York: Knopf, 1928.

Fanning, Leonard. *Foreign Oil and the Free World*. New York: McGraw-Hill, 1954.

Gans, Herbert. *The Levittowners: Ways of Life and Politics in a New Suburban Community*. New York: Pantheon, 1967.

Hamilton, James D. "Oil prices, exhaustible resources, and economic growth." Prepared for *Handbook of Energy and Climate Change*. University of California, January 9, 2012.

Henry, J.T. *The Early and Later History of Petroleum: With Authentic Facts in Regard to Its Development in Western Pennsylvania, the Oil Fields of Europe, and America*. Philadelphia: Rodgers, 1873.

Ise, John. *The United States Oil Policy*. New Haven, CT: Yale University Press, 1926.

Kovarik, Bill. "The whale oil myth." *The Source: Exploring the History of Sustainable Energy* (blog), 2008. At sustainablehistory.wordpress.com/bioenergy/the-whale-oil-myth/.

Knowles, Ruth. *The Greatest Gamblers: The Epic of American Oil Exploration*. New York: McGraw-Hill, 1959.

Ling, Peter. *America and the Automobile: Technology, Reform, and Social Change*. Manchester, U.K.: Manchester University Press, 1990.

Lyman Stewart Guild. At connect.biola.edu/page.aspx?pid=402.

Marcosson, Isaac. *The Black Golconda: The Romance of Petroleum*. New York: Harper, 1923.

McLaurin, John J. *Sketches in Crude-Oil: Some Accidents and Incidents of the Petroleum Development in All Parts of the Globe*. 1896; Franklin, PA, 1902.

McNichol, Dan. *The Roads That Built America: The Incredible Story of the U.S. Interstate System*. Sterling, 2005.

"Oil and the history of Southern California—Question." Interview with Eric Schlosser. *New York Times*, February 22, 2008.

Oklahoma Corporation Commission, Oil and Gas Conservation Division. *A History of Energy*. PowerPoint presentation. At occeweb.com/og/publications.htm.

Olien, Roger, and Diana Davids Olien. *Easy Money: Oil Promoters and Investors in the Jazz Age*. Chapel Hill: University of North Carolina Press, 1990.

Olien, Roger, and Diana Davids Olien. *Oil Booms: Social Change in Five Texas Towns*. Lincoln: University of Nebraska Press, 1982.

Pogue, Joseph E. *The Economics of Petroleum*. New York: Wiley, 1921.

Prindle, David. *Petroleum Politics and the Texas Railroad Commission*. Austin: University of Texas Press, 1981.

Reeve, Sidney Armor. *Energy: Work, Heat and Transformation*. New York: McGraw-Hill, 1909.

Rickover, Hyman G. "Energy resources and our future." Remarks prepared for the Annual Scientific Assembly of the Minnesota State Medical Association, St. Paul, Minnesota, May 14, 1957.

Sabin, Paul. "Home and abroad: The two 'wests' of twentieth-century United States history." *Pacific Historical Review* 66, no. 3 (1997): 305–335.

Solberg, Carl. *Oil Power*. New York: Mason/Charter, 1976.

Stegner, Wallace. *Discovery! The Search for Arabian Oil*. 1971; Vista, CA: Selwa, 2007.

Tarbell, Ida. *The History of the Standard Oil Company*. New York: McClure, Phillips, 1904.

Vance, Rupert. *Human Geography of the South: A Study in Regional Resources and Human Adequacy*. Chapel Hill: University of North Carolina Press, 1932.

Veblen, Thorstein. *The Theory of the Leisure Class: An Economic Study in the Evolution of Institutions*. New York: Macmillan, 1899.

White, Leslie. "Energy and the evolution of culture." *American Anthropologist* 45, no. 3 (1943): 335–356.

White, William. *The Organization Man*. New York: Simon & Shuster, 1956.

CHAPTER FOUR

Angela, Alberto. *A Day in the Life of Ancient Rome: Daily Life, Mysteries and Curiosities*. Translated by Gregory Conti. New York: Europa Editions, 2009.

Bain, H. Foster. "China's coal reserves." *Foreign Affairs*, April 1928.

Dukes, Jeffrey S. "Burning buried sunshine: Human consumption of ancient solar energy." *Climatic Change* 61, nos. 1–2 (2003): 31–44.

Freyre, Gilberto. *The Mansions and the Shanties: The Making of Modern Brazil*. Translated by Harriet de Onils. New York: Knopf, 1963.

Fuller, R. Buckminster. *Nine Chains to the Moon*. Philadelphia: Lippincott, 1938.

Graeber, David. *Debt: The First 5,000 Years*. Brooklyn, NY: Melville House, 2011.

Hodgkinson, Will. "Is slavery the new green energy?" *The Guardian*, December 3, 2009.

Hoffman, Andrew. "Climate change as a cultural and behavioral issue: Addressing barriers and implementing solutions." *Organizational Dynamics* 39 (2010).

Krausse, Joachim, and Claude Lichtenstein, eds. *Your Private Sky: R. Buckminster Fuller, the Art of Design Science*. Translated by Steven Lindberg and Julia Thorsen. Baden: Lars Muller, 2001.

Piller, Ingrid. *American Automobile Names*. PhD Diss., Dresden University, 1995.

Smil, Vaclav. "Energy at the crossroads." Paper presented at the OECD Global Science Forum, Paris, May 17–18, 2006.

Smil, Vaclav. *Why America Is Not a New Rome*. Cambridge, MA: MIT Press, 2010.

Tertzakian, Peter, with Keith Hollihan. *The End of Energy Obesity: Breaking Today's Energy Addiction for a Prosperous and Secure Tomorrow*. Hoboken, NJ: Wiley, 2009.

Ubbelohde, A.R. *Man and Energy*. 1955. Revised ed.; Baltimore, MD: Penguin, 1963.

CHAPTER FIVE

Al-Asi, Taysir. "Overweight and obesity among Kuwait Oil Company employees: A cross-sectional study." *Occupational Medicine* 53 (2003): 431–435.

Berry, Wendell. "The pleasures of eating." In *What Are People For? Essays*. San Francisco: North Point, 1990.

Berry, Wendell. *The Unsettling of America: Culture and Agriculture*. San Francisco: Sierra Club Books, 1977.

Berton, Hal, William Kovarik, and Scott Sklar. *The Forbidden Fuel: A History of Power Alcohol*. Rev. ed. Lincoln: University of Nebraska Press, 2010.

Billard, Jules B., and Blair, James P. "The revolution in American agriculture." *National Geographic* 137, no. 2 (February 1970).

Braudel, Fernand. *Memory and the Mediterranean*. Translated by Siân Reynolds. New York: Knopf, 2001.

Bray, George. "History of obesity." In *Obesity: Science to Practice*, edited by Gareth Williams and Gema Fruhbeck. Chichester, U.K.: Wiley, 2009.

Buncombe, Andrew. "The good life in Havana: Cuba's green revolution." *The Independent*, August 8, 2006.

Canning, Patrick, Ainsley Charles, Sonya Huang, Karen R. Poenske, and Arnold Water. *Energy Use in the U.S. Food System*. Economic Research Report 94. Washington, D.C.: United States Department of Agriculture, Economic Research Service, 2010.

Davis, Adrian, Carolina Valsecchi, and Malcolm Fergusson. *Unfit for Purpose: How Car Use Fuels Climate Change and Obesity*. London: Institute for European Environmental Policy, 2007.

Diamond, Jared. "The worst mistake in the history of the human race." *Discover Magazine* (May 1987): 64–66.

Drewnowski, Adam. "Obesity, diets, and social inequalities." *Nutrition Reviews* 67, Suppl. 1 (2009): S36–S39.

Drewnowski, Adam, and S.E. Specter. "Poverty and obesity: The role of energy density and energy costs." *American Journal of Clinical Nutrition* 79, no. 1 (2004): 6–16.

Eades, Michael R. "Why we get fat" (blog post). Review of Gary Taubes, *Why We Get Fat*. May 6, 2011. At proteinpower.com/drmike/low-carb-library/why-we-get-fat.

Ewing, Ed. "Cuba's organic revolution." *The Guardian*, April 4, 2008.

Food and Agriculture Organization of the United Nations. *"Energy-Smart" Food for People and Climate*. Rome: FAO, 2011.

Fraser, Evan D.G., and Andrew Rimas. *Empires of Food: Feast, Famine, and the Rise and Fall of Civilizations*. New York: Free Press, 2010.

Funes-Monzote, Fernando. *Towards Sustainable Agriculture in Cuba*. 2006. At campus.usal.es/~ehe/Papers/Microsoft%20Word%20-%20Towards%20sustainable%20agriculture%20in%20Cuba%201st%20August[1].pdf.

Gandhi, Mahatma. *Hind Swaraj or Indian Home Rule*. Phoenix, India: International Printing Press, 1909.

Giampietro, Mario, and Kozo Mayumi. *The Biofuel Delusion: The Fallacy of Large-Scale Agro-Biofuel Production*. London: Earthscan, 2009.

Howard, Albert. *An Agricultural Testament*. London: Oxford University Press, 1944.

Jancovici, Jean-Marc. "How much of a slave am I?" *Manicore*, August 2005. At manicore.com/anglais/documentation_a/slaves.html.

Johansson, Kersti, Karin Liljequist, Lars Ohlander, and Kjell Aleklett. "Agriculture as provider of both food and fuel." *AMBIO: A Journal of the Human Environment* 39, no. 2 (2010): 91–99.

Khosruzzaman, Shaikh, M. Ali Asgar, K.M. Rezaur Rahman, and Showkat Akbar. "Energy intensity and productivity in relation to agriculture-Bangladesh perspective." *Journal of Bangladesh Academy of Sciences* 34, no. 1 (2010): 59–70.

Krausmann, Fridolin, Helmut Haberl, Neils B. Schulz, Karl-Heinz Erb, Ekkehard Darge, and Veronika Gaube. "Land-use change and socio-economic metabolism in Austria—Part I: Driving forces of land-use change: 1950–1995." *Land Use Policy* 20, no. 1 (2003): 1–20.

Lang, Susan. "Cornell ecologists' study finds that producing ethanol and biodiesel from corn and other crops is not worth the energy." Cornell University News Service, July 5, 2005. At news.cornell.edu/stories/july05/ethanol.toocostly.ssl.html.

Manning, Richard. "The oil we eat: Following the food chain back to Iraq." *Harper's Magazine*, February 2004.

Marx, Karl, and Friedrich Engels. *The Communist Manifesto*. 1848; London: Penguin, 2002.

Nikiforuk, Andrew. *Pandemonium: Bird Flu, Mad Cow Disease and Other Biological Plagues of the 21st Century*. Toronto: Penguin, 2006.

Odum, Howard T. *Environment, power, and society*. New York: Wiley, 1971.

Patzek, Tad. "Thermodynamics of the corn-ethanol biofuel cycle." *Critical Reviews in Plant Sciences* 23, no. 6 (2004): 519–567.

Philpott, Tom. "An interview with David Pimentel." *Grist*, December 9, 2006. At grist.org/food/philpott2/.

Pimentel, David. "Reducing energy inputs in the agricultural production system." *Monthly Review* 61, no. 3 (2009).

Potter, David M. *People of Plenty: Economic Abundance and the American Character*. Chicago: University of Chicago Press, 1954.

Rathje,William, and Cullen Murphy. *Rubbish! The Archaeology of Garbage*. Tucson: University of Arizona Press, 2001.

Rockström, Johan, et al. (The Resilience Alliance). "Planetary boundaries: Exploring the safe operating space for humanity." *Ecology and Society* 14, no. 2 (2009). At stockholmresilience.org/download/18.8615c78125078c8d3380002197/ES-2009-3180.pdf.

Shiva, Vandana. *Soil Not Oil: Environmental Justice in a Time of Climate Crisis.* Cambridge, MA: South End Press, 2008.
Smil, Vaclav. "Detonator of the population explosion." *Nature*, July 29, 1999.
Smil, Vaclav. "Nitrogen cycle and world food production." *World Agriculture* 2 (2011): 9–11.
Twelve Southerners. *I'll Take My Stand: The South and the Agrarian Tradition.* New York: Harper, 1930.

CHAPTER SIX

Aligica, Paul Dragos. "Julian Simon and the 'limits to growth' neo-Malthusianism." *Electronic Journal of Sustainable Development* 1, no. 3 (2009).
Bardi, Ugo. "The Seneca effect: Why decline is faster than growth." *Cassandra's Legacy*, August 28, 2011. At cassandralegacy.blogspot.ca/2011/08/seneca-effect-origins-of-collapse.html.
Bartlett, Albert. "Reflections on sustainability, population growth and the environment-revisited." *Renewable Resources Journal* 15, no. 4 (1997): 6–23.
Cellier, François. "Ecological footprint, energy consumption, and the looming collapse." *The Oil Drum*, May 16, 2007. At theoildrum.com/node/2534.
Chesterton, G.K. *Eugenics and Other Evils.* London: Cassell, 1922.
Coleman, David, and Robert Rowthorn. "Who's afraid of population decline? A critical examination of its consequences." *Population and Development Review* 37, Suppl. 1 (2011): 217–248.
Ehrlich, Paul. *Population Bomb.* New York: Ballantine, 1968.
Ellul, Jacques. *The Technological Bluff.* Translated by Geoffrey W. Bromily. Grand Rapids, MI: Eerdmans, 1990.
Elton, Charles. "The study of epidemic diseases among wild animals." *Journal of Hygiene* 31, no. 4 (1931): 435–456.
Firth, Niall. "Human race 'will be extinct within 100 years,' claims leading scientist." *Daily Mail*, June 19, 2010.
Fischer, David Hackett. *The Great Wave: Price Revolutions and the Rhythm of History.* New York: Oxford University Press, 1996.
Grantham, Jeremy. "Time to wake up: Days of abundant resources and falling prices are over forever." *GMO Quarterly Letter*, April 2011.
Hubbert, M. King. *Energy Resources.* Washington, D.C.: National Academy of Sciences, National Research Council, 1962.
Hubbert, M. King. "Exponential growth as a transient phenomenon in human history." In *Societal Issues, Scientific Viewpoints*, edited by Margaret A. Strom. New York: American Institute of Physics, 1976.
Kurtz, Stanley. "Demographics and the culture war." *Policy Review* 129 (February 1, 2005).
McNeill, J.R. *Something New under the Sun: An Environmental History of the Twentieth-Century World.* New York: Norton, 2000.
Peters, Gary. "Peak oil and the third demographic transition: A preliminary model." *Our Finite World*, March 28, 2011. At ourfiniteworld.com/2011/03/28/peak-oil-and-the-third-demographic-transition-a-preliminary-model/.

Pimentel, David, and Russell Hopfenberg. "Human population numbers as a function of food supply." *Environment, Development and Sustainability* 3, no. 1 (2001): 1–15.

"The revenge of Malthus: A famous bet recalculated." *The Economist,* August 6, 2011.

Revkin, Andrew, and Vaclav Smil. "9 billion people + 1 planet = ?" Quantum to Cosmos Festival, Perimeter Institute, Waterloo, ON, Canada, October 17, 2009.

Rotella, Carlo. "Can Jeremy Grantham profit from ecological mayhem?" *New York Times Magazine,* August 11, 2011.

Simon, Julian. *The Ultimate Resource.* Princeton, NJ: Princeton University Press, 1981.

Skakkebæk, Niels, et al. "Is human fecundity declining?" *International Journal of Andrology* 29 (2006): 2–11.

World Wildlife Federation. *Living Planet Report.* WWF, 2010. At wwf.panda.org/about_our_earth/all_publications/living_planet_report/2010_lpr/.

Zabel, Graham. "Peak people: The interrelationship between population growth and energy resources." *Energy Bulletin,* April 20, 2009. At energybulletin.net/node/48677.

CHAPTER SEVEN

Batty, Michael, and Stephen Marshall. "The evolution of cities: Geddes, Abercrombie and the New Physicalism." *Town Planning Review* 80, no. 6 (2009): 551–574.

Berry, Wendell. "Out of your car, off your horse." *Atlantic Monthly,* February 1991.

Bettencourt, Luis, and Geoffrey West. "A unified theory of urban living." *Nature* 467 (October 2010): 912–913.

Campanella, Thomas. *The Concrete Dragon: China's Urban Revolution and What It Means for the World.* New York: Princeton Architectural Press, 2008.

Crary, Duncan. *The KunstlerCast: Conversations with James Howard Kunstler.* Gabriola Island, BC: New Society Publishers, 2011.

Dobbs, Richard, and Remes, Jaana. "What's the biggest limit on city growth? (Hint: it's not steel or cement)." *What Matters,* February 1, 2011. At whatmatters.mckinseydigital.com/cities/what-s-the-biggest-limit-on-city-growth-hint-it-s-not-steel-or-cement.

Engel, Katalina, Dorothee Jokiel, Andrea Kraljevic, Martin Geiger, and Kevin Smith. *Big Cities, Big Water, Big Challenges: Water in anUrbanizing World.* Berlin: WWF Germany, August 2011.

"A few words with Vaclav Smil." *Gridlines,* Summer 2011.

Geddes, Patrick. *Cities in Evolution: An Introduction to the Town Planning Movement and to the Study of Civics.* London: Williams & Norgate, 1915.

Girardet, Herbert. "Cities, people, planet." Schumacher Lectures, Liverpool, U.K., April 2000.

Kohr, Léopold. "The eve of 1984." Acceptance speech for the Right Livelihood Award, December 9, 1983. At rightlivelihood.org/kohr_speech.html.

Lehrer, Jonah. "A physicist solves the city." *New York Times Magazine*, December 17, 2010.
Lemann, Nicholas. "Has the celebration of cities gone too far?" *The New Yorker*, June 27, 2011.
Lerup, Lars. *After the City*. Cambridge, MA: MIT Press, 2000.
Lerup, Lars. "Toxic ecology: The struggle between nature and culture in the suburban megacity." Megacities Lecture, Amsterdam, November 17, 2005. At megacities.nl/?page_id=82.
Mumford, Lewis. *The City in History: Its Origins, Its Transformations, and Its Prospects*. New York: Harcourt, Brace, 1961.
Mumford, Lewis. *The Culture of Cities*. 1938; New York: Harcourt Brace Jovanovich, 1970.
Newman, Peter. "Sustainability and cities: Extending the metabolism model." *Landscape and Urban Planning* 44, no. 4 (1999): 219–226.
Pedersen, Martin C. "The Chinese century." Interview with Thomas Campanella. *Metropolis Magazine*, September 2008.
Phdungslip, Aumnad. *Energy Analysis for Sustainable Mega-Cities*. Licentiate thesis, School of Industrial Engineering and Management, Stockholm, 2006.
Reeve, Sidney A. "Congestion in cities." *The Geographical Review* 3, no. 4 (1917).
Schumacher, E.F. *Small Is Beautiful: Economics as if People Mattered*. New York: Harper & Row, 1973.
Sheehan, Paul. "Life's a bitumen nightmare as cities get hotter than hell." *Sidney Herald*, February 15, 2010.
UN-Habitat. *State of the World's Cities 2008/2009: Harmonious Cities*. UN-Habitat, 2008.
Watts, Jonathan. *When a Billion Chinese Jump: How China Will Save Mankind or Destroy It*. London: Faber and Faber, 2010.
Williams, Mike. "Metropolitan man: Departing Architecture dean Lars Lerup takes a hard look at Houston's future." *Rice News*, July 10, 2009. At news.rice.edu/2009/07/10/departing-architecture-dean-lars-lerup-takes-a-hard-look-at-houstons-future/.

CHAPTER EIGHT

Aleklett, Kjell. "Peak oil and the evolving strategies of oil importing and exporting countries." In OECD International Transport Forum, Round Table 139, *Oil Dependence: Is Transport Running Out of Affordable Fuel?* OECD, 2008.
Aleklett, Kjell, Mikael Höök, Kristofer Jakobsson, Michael Lardelli, Simon Snowden, and Bengt Söderbergh. "The peak of the oil age: Analyzing the world oil production Reference Scenario in World Energy Outlook 2008." *Energy Policy* 38, no. 3 (2010): 1398–1414.
Ayres, Robert. "Energy intensity, efficiency and economics." Lecture, IMF Research Department, December 7, 2010.
Ayres, Robert U., and Benjamin Warr. "Accounting for growth: The role of physical work." In *Advances in Energy Studies*, edited by Sergio Ulgiati. Padova, Italy: SGEditoriali, 2003; *Structural Change and Economic Dynamics* 16 (2005): 181–209.

Beaudreau, Bernard C. "Engineering and economic growth." October 2001. At papers.econ.mpg.de/evo/Conference_papers/Production/Beaudreau.pdf.

Bradley, Robert, Jr. "Dear peak oilers: Please consider Erick Zimmermann's 'functional theory' of mineral resources." *MasterResource* (blog), October 22, 2010. At masterresource.org/2010/10/dear-peak-oilers-zimmermanns-functional-theory/.

Bradley, Robert L., Jr., and Richard W. Fulmer. *Energy: The Master Resource.* Dubuque, IA: Kendall, 2004.

Cleveland, Cutler J . "Biophysical economics: From physiocracy to ecological economics and industrial ecology." In *Bioeconomics and Sustainability: Essays in Honor of Nicholas Georgescu-Roegen*, edited by Kozo Mayumi and John Gowdy. Cheltenham, U.K.: Edward Elgar, 1999.

Cowen, Tyler. *The Great Stagnation: How America Ate All the Low-Hanging Fruit of Modern History, Got Sick, and Will (Eventually) Feel Better.* New York: Penguin, 2011.

Frey, Donald E. *America's Economic Moralists: A History of Rival Ethics and Economics.* Albany: State University of New York Press, 2009.

Geddes, Patrick. *John Ruskin: Economist.* Edinburgh: William Brown, 1884.

Georgescu-Roegen, Nicholas. "Energy and economic myths." *Southern Economic Journal* 41, no. 3 (1975).

Georgescu-Roegen, Nicholas. "The entropy law and the economic process in retrospect." *Eastern Economic Journal* 12, no. 1 (1986): 3–25.

Gowdy, John, and Susan Mesner. "The Evolution of Georgescu-Roegen's Bioeconomics." *Review of Social Economy* 41, no. 2 (1998): 136–156.

Hall, Charles. "Exchange on the difference between biophyscial and ecological economics." Email to A., March 3, 2009. At biophysicalecon.blogspot.ca.

Hall, Charles A., and Kent Klitgaard. "The need for a new, biophysical-based paradigm in economics for the second half of the age of oil." *International Journal of Transdisciplinary Research* 1, no. 1 (2006): 4–22.

Hall, Charles A., and Joe-Young Ko. "Energy and international development: A systems approach to economic development," In *Proceedings of IV Biennial International Workshop "Advances in Energy Studies,"* edited by Enrique Ortega and Sergio Ulgiati. Unicamp, Campinas, São Paolo, Brazil, June 16–19, 2004.

Hamilton, James D. "Historical oil shocks." Prepared for *Handbook of Major Events in Economic History.* University of California, San Diego, February 1, 2011.

Hamilton, James D. "Oil prices, exhaustible resources, and economic growth." Prepared for *Handbook of Energy and Climate Change.* University of California, January 9, 2012.

Hobson, J.A. *John Ruskin: Social Reformer.* Boston: Dana Estes, 1898.

Jevons, W. Stanley. *The Coal Question: An Inquiry Concerning the Progress of the Nation, and the Probable Exhaustion of Our Coal-Mines.* 3rd ed. 1866; New York: Augustus M. Kelley, 1965.

Lebow, Victor. "Price competition in 1955." *Journal of Retailing* 31, no. 1 (1955).

Leontief, Wassily W. Letter. *Science*, July 9, 1982.

Martinez-Alier, Juan. *Energy-Related Issues in Early Economic Literature.* Ottawa: Energy Research Group, March 1986.

Marx, Karl, and Friedrich Engels. *The Communist Manifesto.* 1848; London: Penguin, 2002.

MasterResource: A Free-Market Energy Blog. At masterresource.org/.

McNeill, J.R. *Something New under the Sun: An Environmental History of the Twentieth-Century World.* New York: Norton, 2000. Orlov, Dmitry. "Closing the 'collapse gap': The USSR was better prepared for collapse than the U.S." *Energy Bulletin,* December 4, 2006.

"Powering the economic growth engine." INSEAD, December 28, 2010. At knowledge.insead.edu/TheEconomicGrowthEngine090716.cfm.

Reynolds, Douglas. "Peak oil and the fall of the Soviet Union: Lessons of the collapse." *The Oil Drum,* May 21, 2011. Available at theoildrum.com/node/7878.

Ruskin, John. *The Communism of John Ruskin; or, "Unto This Last": Two Lectures from "The Crown of Wild Olive"; and Selections from "Fors Clavigera."* Edited by W.D.P. Bliss. New York: Humboldt, 1891.

Ruskin, John. *The Seven Lamps of Architecture.* 1849. Orpington; London: George Allen, 1889.

Samuelson, Paul, and William Nordhaus, *Economics.* 18th ed. New York: McGraw-Hill, 2004.

Simon, Julian. *The Ultimate Resource.* Princeton, NJ: Princeton University Press, 1981.

Smith, Adam. *An Inquiry into the Nature and Causes of the Wealth of Nations.* 2 vols. 1776; Oxford University Press, 1976; rpt. Indianapolis, IN: Liberty Fund, 1982.

Soddy, Frederick. *Cartesian Economics: The Bearing of Physical Science upon State Stewardship.* London: Hendersons, 1921. Soddy, Frederick. *Matter and Energy.* London: Williams & Norgate, 1912.

Soddy, Frederick. *The Role of Money: What It Should Be, Contrasted with What It Has Become.* London: George Routledge, 1934.

Soddy, Frederick. *Wealth, Virtual Wealth and Debt: The Solution of the Economic Paradox.* London: Allen & Unwin, 1926.

Solow, Robert. "An almost practical step toward sustainability." In National Research Council, Commission on Geosciences, Environment, and Resources and Commission on Behavioral and Social Sciences and Education, *Assigning Economic Value to Natural Resources.* Washington, D.C.: National Academy Press, 1994.

Solow, Robert. "The economics of resources or the resources of economics." *American Economics Review* 64 (1974): 1–14.

Spiegel, Henry. *The Growth of Economic Thought.* 2nd ed. Durham, NC: Duke University Press, 1983.

Taleb, Nicholas Nassim. *The Black Swan: The Impact of the Highly Improbable.* 2nd ed. New York: Random House, 2010.

Zimmermann, Erich. *World Resources and Industries: A Functional Appraisal of the Availability of Agricultural and Industrial Resources.* New York: Harper, 1933.

CHAPTER NINE

Angelica, Amara. "A limitless power source for the indefinite future." *Kurzweil Accelerating Intelligence Blog*, November 11, 2011. At kurzweilai.net/a-limitless-power-source-for-the-indefinite-future.

Bardi, Ugo. "Peak research." *Cassandra's Legacy*, July 22, 2011. At cassandralegacy.blogspot.ca/2011/07/peak-research.html.

Carpenter, Edward. *Civilisation: Its Cause and Cure, and Other Essays*. London: Swan Sonnenschein, 1897.

Carrel, Alexis. *Man, the Unknown*. New York: Harper, 1935.

"China's phony science." *The New Atlantis* (Summer 2006): 103–106.

Coy, Peter. "The other U.S. energy crisis: Lack of R&D." *Bloomberg Businessweek*, June 17, 2010. At businessweek.com/magazine/content/10_26/b4184029812114.htm.

De Decker, Kris. "The monster footprint of digital technology." *Low-tech Magazine*, June 16, 2009. At lowtechmagazine.com/2009/06/embodied-energy-of-digital-technology.html.

Ellul, Jacques. *Perspectives on Our Age: Jacques Ellul Speaks on His Life and Work*. Edited by William H. Vanderburg. Translated by Joachim Neugroschel. Originally broadcast on *Ideas*, CBC Radio. Toronto: Canadian Broadcasting Corporation, 1981.

Ellul, Jacques. *The Technological Bluff*. Translated by Geoffrey W. Bromily. Grand Rapids, MI: Eerdmans, 1990.

Ellul, Jacques. *The Technological Society*. Translated by John Wilkinson. New York: Knopf, 1964.

Epstein, Richard. *Overdose: How Excessive Government Regulation Stifles Pharmaceutical Innovation*. New Haven, CT: Yale University Press, 2006.

Glass, Bentley. "On scientific progress and its limits." *Quarterly Review of Biology* 54, no. 4 (1979): 417–419.

Gutowski, Thomas G., Matthew S. Branham, Jeffrey B. Dahmus, Alissa J. Jones, and Alexandre Thiriez. "Thermodynamic analysis of resources used in manufacturing processes." *Environmental Science & Technology* 43, no. 5 (2009): 1584–1590.

Hansson, Anders, and Mårten Bryngelsson. "Expert opinions on carbon dioxide capture and storage—A framing of uncertainties and possibilities." *Energy Policy* 37, no. 6 (2009): 2273–2282.

Hays, Jeffrey. "Academic misconduct in China." FactsandDetails.Com. Accessed July 2011 at factsanddetails.com/china.php?itemid=1651&catid=13&subcatid=82.

Huebner, Jonathan. "A possible declining trend for worldwide innovation." *Technological Forecasting & Social Change* 72 (2005): 980–986.

Huxley, Aldous. *Science, Liberty and Peace*. 1946; London: Chatto & Windus, 1950.

Javitz, Harold, Teresa Grimes, Derek Hill, Alan Rapoport, Robert Bell, Ron Fesco, and Rolf Lehming. *U.S. Adademic Scientific Publishing*. Working Paper. National Science Foundation, 2010.

Jones, Steve. "One gene will not reveal all life's secrets." *The Telegraph*, April 20, 2009.

Liao, Matthew. "Human Engineering and Climate Change." February 2, 2012. *Ethics, Policy and the Environment*, forthcoming.

Liebig, Justus von. *Familiar Letters on Chemistry, in Its Relations to Physiology, Dietetics, Agriculture, Commerce, and Political Economy*. 3rd ed. London: Taylor, Walton, & Maberly, 1851.

Mankins, John C., ed. *Space Solar Power: The First International Assessment of Space Solar Power: Opportunities and Potential Pathways Forward*. International Academy of Astronautics. Toronto: Space Canada, 2011.

Martin, Brian. "Scientific fraud and the power structure of science." *Prometheus* 10, no. 1 (1992): 83–98.

Mervis, Jeffrey. "U.S. output flattens, and NSF wonders why." *Science*, August 3, 2007.

Naik, Gautam. "Mistakes in scientific studies surge." *Wall Street Journal*, August 10, 2011.

Nemet, Gregory F., and Daniel M. Kammen. "U.S. energy research and development: Declining investment, increasing need, and the feasibility of expansion." *Energy Policy* 35 (2007): 746–755.

Price, Derek J. de Solla. *Little Science, Big Science—and Beyond*. First edition published as *Little Science, Big Science*, 1963. New York: Columbia University Press, 1986.

Rescher, Nicholas. *The Limits of Science*. Rev. ed. Pittsburgh, PA: University of Pittsburgh Press, 1999.

Retraction Watch. At retractionwatch.wordpress.com.

"Scientific fraud: Action needed in China." *The Lancet*, January 9, 2010.

Sclove, Richard E. "From alchemy to atomic war: Frederick Soddy's 'technology assessment' of atomic energy, 1900–1915." *Science, Technology & Human Values* 14, no. 2 (1989): 163–194.

Smil, Vaclav. "Energy at the crossroads." Paper presented at the OECD Global Science Forum, Paris, May 17–18, 2006.

Smil, Vaclav. "Long-range energy forecasts are no more than fairy tales," *Nature*, May 8, 2008.

Soddy, Frederick. *Science and Life: Aberdeen Addresses*. London: J. Murray, 1920.

Spreng, Daniel, Gregg Marland, and Alvin M. Weinberg. "CO_2 capture and storage: Another Faustian bargain?" *Energy Policy* 35, no. 2 (2007): 850–854.

Steen, R. Grant. "Retractions in the scientific literature: Is the incidence of research fraud increasing?" *Journal of Medical Ethics* 37 (2011): 249–253.

Tainter, Joseph. "Problem solving: Complexity, history, sustainability, population and environment." *Population and Environment* 22, no. 1 (2000): 3–41.

Tainter, Joseph, T.F.H. Allen, Amanda Little, and Thomas W. Hoekstra. "Resource transitions and energy gain: Contexts of organization." *Conservation Ecology* 7, no. 3 (2003).

Unruh, Gregory. "Understanding carbon lock-in." *Energy Policy* 28, no. 12 (2000): 817–830.

Verleger, Philip K., Jr. "Forty years of folly: The failure of U.S. energy policy." *International Economy*, Winter 2011.

CHAPTER TEN

Berry, Jason. "BP storm: Tulane prof Oliver Houck warned for decades of peril of lax energy regulations." *Politics Daily*, June 6, 2010. At politicsdaily.com /2010/06/13/bp-storm-tulane-prof-oliver-houck-warned-for-decades-of-peril-o/.

Blanchard, Christopher. *The Islamic Traditions of the Wahhabism and Salafiyya*. Congressional Research Service Report for Congress, Order Code RS21695. Washington, D.C.: Library of Congress, January 24, 2008.

Bower, Tom. *Oil: Money, Politics, and Power in the 21st Century*. London: Harper, 2010.

Burrough, Bryan. *The Big Rich: The Rise and Fall of the Greatest Texas Oil Fortunes*. New York: Penguin, 2009.

Cornwall Alliance for the Stewardship of Creation. At cornwallalliance.org.

Curry, Judith. "Understanding conservative religious resistance to climate science." Interview with David Gushee. *Climate Etc* (blog), December 20, 2010. At judithcurry.com/2010/12/20understanding-conservative-religious-resistance-to-climate-science/.

Fineberg, Richard. "Commentary: An introduction to petropolitics." *Fineberg Research Archives*, August 2004. At finebergresearch.com/archives/arcpetropolitics.html.

Goldberg, Ellis, Erik Wibbels, and Eric Mvukiyehe. "Lessons from strange cases: Democracy, development and the resource curse in the United States." *Comparative Political Studies* 4, no. 4/5 (2008): 477–514.

Gorshkov, Victor G., Anastassia M. Makarieva, and Bai-Lian Li. "Comprehending ecological and economic sustainability: Comparative analysis of stability principles in the biosphere and free market economy." *Annals of the New York Academy of Sciences* 1195 (2010): E1–E18.

Green, Joshua. "The tragedy of Sarah Palin." *The Atlantic*, June 2011.

Gushee, David. *The Future of Faith in American Politics: The Public Witness of the Evangelical Center*. Waco, TX: Baylor University Press, 2008.

Homans, Charles. "RIP Ted Stevens, architect of the Alaskan petrostate." *Foreign Policy*, August 10, 2010. At oilandglory.foreignpolicy.com/posts/2010/08/10/rip_ted_stevens_architect_of_the_alaskan_petrostate.

Juhasz, Antonia. *The Tyranny of Oil: The World's Most Powerful Industry—and What We Must Do to Stop It*. New York: William Morrow, 2008.

Karl, Terry Lynn. *Democracy over a Barrel: Oil, Regime Change and War*. CSD Working Papers 8, no. 7. Center for the Study of Democracy, University of California-Irvine, July 7, 2008.

Karl, Terry Lynn. *Oil-Led Development: Social, Political, and Economic Consequences*. CDDRL Working Papers no. 80. Center on Democracy, Development, and the Rule of Law, Stanford University, January 2007.

Karl, Terry Lynn. *The Paradox of Plenty: Oil Booms and Petro-States*. Berkeley: University of California Press, 1997.

Karl, Terry Lynn. "The perils of the petro-state: Reflections on the paradox of plenty." *Journal of International Affairs* 53, no. 1 (1999).

Kelley, Wayne, and Richard Bishop. *Global Oil Trade: The Relationship between Wealth Transfer and Giant Fields*. Houston: RSK Limited, September 20, 2010. At rskuklimited.com/news/?p=156.

Kruse, Kevin. "For God so loved the 1 percent." *New York Times,* January 17, 2012.
Martinez, Ibsen. "The curse of the petro-state: The example of Venezuela." *Library of Economics and Liberty,* September 5, 2005. At econlib.org/library/Columns/y2005/Martinezpetro.html.
Mitchell, Timothy. *Carbon Democracy: Political Power in the Age of Oil.* New York: Verso, 2011.
Mufson, Steven. "America's petro-state: Louisiana has paid a steep price for its bargain with the oil industry." *Washington Post,* July 24, 2010.
Orlov, Dimitry. "Closing the 'Collapse Gap': The USSR was better prepared for collapse than the U.S." *Energy Bulletin,* December 4, 2006. At energybulletin.net/node/23259.
O'Rourke, Dara, and Sarah Connolly. "Just oil? The distribution of environmental and social impacts of oil production and consumption." *Annual Review of Environment and Resources* 28 (2003): 587–617.
Ross, Michael. *The Oil Curse: How Petroleum Wealth Shapes the Development of Nations.* Princeton, NJ: Princeton University Press, 2012.
Ross, Michael. "Oil, Islam, and women." *American Political Science Review* 102, no. 1 (2008): 107–123.
Ryggvik, Helge. *The Norwegian Oil Experience: A Toolbox for Managing Resources?* Translated by Laurence Cox. Oslo: Centre for Technology, Innovation and Culture, University of Oslo, 2010. At dublinopinion.com/downloads/Norwegian_Oil_Experience_ILR.pdf.
Sætre, Simen. *Petromania.* Oslo: J.M. Stenersens Forlag, 2009.
Sala-i-Martin, Xavier, and Arvind Subramanian. *Addressing the Natural Resource Curse: An Illustration from Nigeria.* IMF Working Paper. Washington D.C.: International Monetary Fund, July 2003.
Sandu, Martin. "The Iraqi who saved Norway from oil." *Financial Times,* August 29, 2009. Schatz, Sayre P. "Pirate capitalism and the inert economy of Nigeria." *Journal of Modern African Studies* 22, no. 1 (1984): 44–57.
Signer, Michael. *Demagogue: The Fight to Save Democracy from Its Worst Enemies.* New York: Palgrave Macmillan, 2009.
Stern, Roger. "United States cost of military force projection in the Persian Gulf, 1976–2007." *Energy Policy* 38 (2010): 2816–2825.
Thompson, Chuck. *Smile When You're Lying: Confessions of a Rogue Travel Writer.* New York: Holt, 2007.
Vulliamy, Ed. "Dark heart of the American dream." *The Observer Magazine,* June 16, 2002.
Watts, Michael. "Oil, development, and the politics of the bottom billion." *Macalester International* 24 (2009). At digitalcommons.macalester.edu/macintl/vol24/iss1/11.
White, Richard D., Jr. *Kingfish: The Reign of Huey P. Long.* New York: Random House, 2006.

CHAPTER ELEVEN

Bardi, Ugo. "The renewable revolution—II." *Cassandra's Legacy* (blog), October 3, 2011. At cassandralegacy.blogspot.ca/2011/10/renewable-revolution-ii.html.

Beal, Colin, Robert E. Hebner, Michael E. Webber, Rodney S. Ruoff, and A. Frank Seibert. "The energy return on investment for algal biocrude: Results for a research production facility." *BioEnergy Research* (July 2011).

Berndt, Ernst R. *From Technocracy to Net Energy Analysis: Engineers, Economists and Recurring Energy Theories of Value*. Studies in Energy and the American Economy Discussion Paper 11.MIT-EL 81-065WP. 1982; in *Progress in Natural Resource Economics*, edited by Anthony Scott, John F. Helliwell, Tracy R. Lewis, and Philip A. Neher. Oxford: Clarendon, 1985.

Chamberlain, Alexander Francis. *The Child: A Study in the Evolution of Man*. London: Walter Scott, 1900.Cleveland, Cutler J. "Biophysical economics: Historical perspectives and current recent trends." *Ecological Modelling* 38, nos. 1–2 (1987): 47–73.

Cleveland, Cutler J. "National resource scarcity and economic growth revisited: Economic and biophysical perspectives. In *Ecological Economics: The Science and Management of Sustainability*, edited by Robert Costanza. New York: Columbia University Press, 1991.

Emerson, Ralph Waldo. "*New England*, lecture II: The trade of New England." January 17, 1843. In *The Later Lectures of Ralph Waldo Emerson, 1843–1871*, vol. 1. Athens: University of Georgia Press, 2010.

Groos, Karl. "The surplus energy theory of play." In *The Play of Animals*, translated by Elizabeth L. Baldwin. New York: Appleton, 1898.

Gupta, Ajay K., and Charles A.S. Hall. "A review of the past and current state of EROI data." *Sustainability* 3 (2011): 1796–1809.

Hall, Charles A.S., guest editor. Special Issue: New Studies in EROI (Energy Return on Investment). *Sustainability* (2011). At mdpi.com/journal/sustainability/special_issues/New_Studies_EROI/.

Hall, Charles A.S., Stephen Balogh, and David J. Murphy. "What is the minimum EROI that a sustainable society must have?" *Energies* 2, no. 1 (2009): 25–47.

Hall, Charles A.S., and Cutler J. Cleveland. "Petroleum drilling and production in the United States: Yield per effort and net energy analysis." *Science* 211 (1981): 576–579.

Höök, Mikael, Junchen Li, Kersti Johansson, and Simon Snowden. "Growth rates of global energy systems and future outlooks." *Natural Resources Research* 21, no. 1 (2012): 23–41.

Pauly, Daniel. "Aquacalypse now: The end of fish." *New Republic*, September 28, 2009.

Pauly, Daniel. "Beyond duplicity and ignorance in global fisheries." *Scientia Marina* 73, no. 2 (2009): 215–224.

Pauly, Daniel. "Toward a conservation ethic for the sea: Steps in a personal and intellectual odyssey." *Bulletin of Marine Science* 87, no. 2 (2011): 165–175.

Pauly, Daniel. Jackie Alder, Elena Bennett, Villy Christensen, Peter Tyedmers, and Reg Watson. "The future for fisheries." *Science* 302 (2003): 1359–1361.

Pauly, Daniel, and Rainer Froese. "Comments on FAO's State of Fisheries and Aquaculture, or 'SOFIA 2010.'" *Marine Policy* 36 (2012): 746–752.

Pauly, Daniel, et al. "The Future For Fisheries." *Science* 302 (2003): 1359–61.

Sahlins, Marshall. "Notes on the original affluent society." In *Man the Hunter*, edited by Richard B. Lee and Irven DeVore. Chicago: Aldine, 1969.

Schiller, Friedrich. *Aesthetical and Philosophical Essays*. Project Gutenberg, 2006. Tyedmers, Peter. "Fisheries and energy use." *Encyclopedia of Energy*, vol. 2. Elsevier, 2004.

CHAPTER TWELVE

Altman, Daniel. "United States of narcissism." *The Daily Beast*, July 17, 2011. At thedailybeast.com/newsweek/2011/07/17/narcissism-is-on-the-rise-in-america.html.

Barrie, D.B., and D.B. Kirk-Davidoff. "Weather response to a large wind turbine array." *Atmospheric Chemistry and Physics* 10 (2010): 769–775.

Bok, Derek. *The Politics of Happiness: What Government Can Learn from the New Research on Well-Being*. Princeton, NJ: Princeton University Press, 2010.

Brown, Brené. "The power of vulnerability." TED talk, June 2010. At ted.com/talks/brene_brown_on_vulnerability.html.

Fridley, David. *Nine Challenges of Alternative Energy*. Santa Rosa, CA: Post Carbon Institute, 2010.

Gandhi, Mahatma. *Hind Swaraj or Indian Home Rule*. Phoenix, India: International Printing Press, 1909.

Gourevitch, Philip. "No Exit: Can Nicolas Sarkozy and France survive the European crisis?" *The New Yorker*, December 12, 2011.

Jevons, W. Stanley. *The Coal Question: An Inquiry Concerning the Progress of the Nation, and the Probable Exhaustion of Our Coal-Mines*. 3rd ed. 1866; New York: Augustus M. Kelley, 1965.

Lewis, C.S. *The Abolition of Man*. London: Oxford University Press, 1943.

Lotka, Alfred. "Contribution to the energetics of evolution." *Proceedings of the National Academy of Sciences of the United States of America* 8, no. 6 (1922): 147–151.

Makarieva, Anastassia, Victor G. Gorshkov, and Bai-Lin Li. "Energy budget of the biosphere and civilization: Rethinking environmental security of global renewable and non-renewable resources." *Ecological Complexity* 5, no. 4 (2008): 281–288.

Millard-Ball, Adam, and Lee Schipper. "Are we reaching a plateau or 'peak' travel?" Paper submitted to the 2010 Transportation Research Board Annual Meeting. Stanford: Global Metropolitan Studies, 2009.

Nieli, Russell. "Critic of the sensate culture: Rediscovering the genius of Pitirim Sorokin." *Political Science Reviewer* 35, no. 1 (2006).

Parker, George. "No death, no taxes: The libertarian futurism of a Silicon Valley billionaire." *The New Yorker*, November 28, 2011.

Penty, Arthur. *Old Worlds for New: A Study of the Post-Industrial State*. London: Allen and Unwin, 1917.

Putnam, Robert. *Bowling Alone: The Collapse and Revival of American Community*. New York: Simon & Schuster, 2000.

Rosenbloom, Stephanie. "But will it make you happy?" *New York Times*, August 7, 2010.

Smil, Vaclav. "Global energy: The latest infatuations." *American Scientist* 99 (May–June 2011): 212–219.

Smil, Vaclav. "A Hummer in every driveway." *Foreign Policy*, November 2011.
Smil, Vaclav. "Science, energy, ethics, and civilization." In *Visions of Discovery: New Light on Physics, Cosmology, and Consciousness*, edited by Raymond Y. Chiao, Marvin L. Cohen, Anthony J. Leggett, William D. Phillips, and Charles L. Harper Jr. Cambridge: Cambridge University Press, 2010.
Sorokin, Pitirim. *Social and Cultural Dynamics*. 4 vols.; New York: American Book Company, 1937–41. Revised and abridged; New Brunswick, NJ: Transaction Books, 1957.
Sorokin, Pitirim A. *Society, Culture, and Personality: Their Structure and Dynamics*. New York: Harper, 1947.
Sorokin, Pitirim. *The Ways and Power of Love: Types, Factors, and Techniques of Moral Transformation*. Boston: Beacon, 1954; Philadelphia: Templeton Foundation Press, 2002.
Wang, C., and R.G. Prinn. "Potential climatic impacts and reliability of very large-scale wind farms." *Atmospheric Chemistry and Physics* 10 (2010): 2053–2061.

CHAPTER THIRTEEN

Citizens' Nuclear Information Center (CNIC). At cnic.jp.
Fackler, Martin. "Japan goes from dynamic to disheartened." *New York Times*, October 16, 2010.
Kerr, Alex. *Dogs and Demons: Tales from the Dark Side of Japan*. New York: Hill and Wang, 2001.
Kunstler, James Howard. "Jim Kunstler's forecast 2011." At kunstler.com/Mags_Forecast2011.php. http://www.kunstler.com/Mags_Forecast2011.php.
Matanle, Peter, and Anthony Rausch with the Shrinking Regions Research Group. *Japan's Shrinking Regions in the 21st Century: Contemporary Responses to Depopulation and Socioeconomic Decline*. Amherst, MA: Cambria Press, 2011.
Ōe, Kenzaburō. "History repeats." *The New Yorker*, March 28, 2011.
Osnos, Evan. "The Fallout: Seven Months Later: Japan's Nuclear Predicament." *The New Yorker*, October 17, 2011.
Peterson, Britt. "Land of disaster." *Foreign Policy*, March 14, 2011.
Smil, Vaclav. "Japan's crisis: Context and outlook." *The American Magazine*, April 16, 2011. At american.com/archive/2011/april/japan2019s-crisis-context-and-outlook.
Smil, Vaclav. "Light behind the fall: Japan's electricity consumption, the environment, and economic growth." *The Asia-Pacific Journal: Japan Focus*, April 2, 2007. At japanfocus.org/-Vaclav-Smil/2394.
Smil, Vaclav. "The unprecedented shift in Japan's population: Numbers, age, and prospects." *The Asia-Pacific Journal: Japan Focus*, May 1, 2007. At japanfocus.org/-Vaclav-Smil/2411.
Smil, Vaclav. *Why America Is Not a New Rome*. Cambridge, MA: MIT Press, 2010.
Tsutsui, William. "Framing twentieth-century Japan: A top-ten list." *About Japan: A Teacher's Resource*, September 7, 2007. At aboutjapan.japansociety.org/content.cfm/framing_twentieth-century_japan_a_top-ten_list.

EPILOGUE

MacIntyre, Alasdair. *After Virtue: A Study in Moral Theory*. 3rd edition. Notre Dame, IN: University of Notre Dame Press, 2007.

Mumford, Lewis. *The Myth of the Machine: Technics and Human Development*. London: Secker & Warburg, 1967.

Mumford, Lewis. *Technics and Civilization*. New York: Harcourt, Brace, 1934.

"The Rule of St. Benedict." c. 530. Translated by Ernest F. Henderson, in *Select Historical Documents of the Middle Ages*, London: George Bell, 1910. Extracts, *Internet Medieval Source Book*, at fordham.edu/halsall/source/rul-benedict.asp.

Williams, Michael. *Deforesting the Earth: From Prehistory to Global Crisis*. Chicago: University of Chicago Press, 2003.

Woods, Thomas E., Jr. *How the Catholic Church Built Western Civilization*. Washington, D.C.: Regnery, 2005.

Acknowledgments

. . . . .

THIS BOOK PRESENTS a rough sketch of several radical ideas about energy and slavery. It draws heavily on the intellectual legacy of Fred Cottrell, Ivan Illich, Frederick Soddy, John Ruskin, Léopold Kohr, Earl Cook, and Adam Hochschild. Critical research by political scientist Terry Lynn Karl on the corrupt and dysfunctional nature of petrostates shaped the tenor of this inquiry. (Unfortunately, Karl was too ill for an interview.) The U.S. ecologist Charles Hall generously shared an early manuscript of his book as well as many important ideas on energy returns. The voluminous works of Vaclav Smil, perhaps North America's greatest energy thinker (and certainly the best read) informs many chapters of this book. Dave Hughes, one of Canada's most clear-eyed energy thinkers and a member of the Post Carbon Institute, provided helpful facts, corrections, and suggestions. Daniel Pauly, fish biologist extraordinaire, lucidly laid out the similarities between unconventional fuels and creatures at the bottom of the marine food web. Last but not least, one-party rule in the petrostate of Alberta for forty-one years has taught me how oil can enslave and neutralize a citizenry as totally as a nineteenth-century plantation economy.

Every discussion about energy consumption is a moral one because it involves some form of slavery. As such, and with no apologies on the matter, three prominent Christian writers have informed my historical essay and polemic: G.K Chesterton, Ivan Illich, and Pitirim Sorokin.

I have written about the oil and gas industry for a variety of magazines and newspapers for more than two decades. My stories about shale gas, conventional oil, coal bed methane (CBM), and bitumen production laid the foundation for this book and its uncomfortable conclusions. Watching the $200 billion tar sands project undermine Canada's character and politics has convinced me that business as usual is over and the oil machine has grown wildly corrupt.

As a provocative sketch on oil and its economic enslavements, this book is admittedly incomplete. The subject begs more inquiry, debate, and studies. An entire chapter could have been devoted to oil's overwhelming monopoly on transportation fuels and automobile economies. Another chapter could have examined how high energy spending has subverted education and turned it into a standardized training vehicle for high-energy living. The impact of armies of inanimate slaves on the hearts and souls of human beings deserves much more reflection. Another chapter could have documented the way hydrocarbon pollution has changed human health by replacing infectious diseases with plagues of cancer, diabetes, and autoimmune ailments. Yet another chapter could have tracked how the availability and quality of energy has changed gender roles and influenced sexual revolutions over the centuries. And the evolution of fundamentalist fossil-fuel cults could fill another book altogether.

Barbara Pulling, my hardworking editor for three consecutive books, ably transformed a large and cumbersome manuscript into a scrappy reading experience—and in an outrageous time frame. I warmly thank her for her ceaseless energy and sound judgment. Stephanie Fysh did a yeoman's job on the copyediting. The team at Greystone Books (Rob Sanders, Emiko Morita, Corina Eberle, and Carra Simpson) performed miracles again.

My wife, Doreen Docherty, whose blood is strong and whose heart dreams in the Hebrides, has sustained our family over the years more than my meager writings have. She is, as Robbie Burns might say, the "guid" author, both merry and wise.

All errors are my responsibility. When I write a book without any mistakes (and I continue trying), I'll be dead.

Readers interested in the evolution of my energy thinking should consult *The Tyee*, Canada's best independent paper: thetyee.ca. I thank *Tyee* editor David Beers and the paper's engaged readers for their continued support and encouragement.

I can be reached at andrew@andrewnikiforuk.com.

Index

. . . . .

THE DAVID SUZUKI FOUNDATION

THE DAVID SUZUKI FOUNDATION works through science and education to protect the diversity of nature and our quality of life, now and for the future.

With a goal of achieving sustainability within a generation, the Foundation collaborates with scientists, business and industry, academia, government, and non-governmental organizations. We seek the best research to provide innovative solutions that will help build a clean, competitive economy that does not threaten the natural services that support all life.

The Foundation is a federally registered independent charity that is supported with the help of over 50,000 individual donors across Canada and around the world.

We invite you to become a member. For more information on how you can support our work, please contact us:

The David Suzuki Foundation
219–2211 West 4th Avenue
Vancouver, BC Canada V6K 4S2
www.davidsuzuki.org
contact@davidsuzuki.org
Tel: 604-732-4228
Fax: 604-732-0752

Checks can be made payable to the David Suzuki Foundation. All donations are tax-deductible.

Canadian charitable registration: (BN) 12775 6716 RR0001
U.S. charitable registration: #94-3204049